AF247505

BLOCK DESIGNS

Analysis, Combinatorics
and Applications

Series on
Applied Mathematics
Volume 17

BLOCK DESIGNS

Analysis, Combinatorics and Applications

Damaraju Raghavarao

Temple University, USA

Lakshmi V. Padgett

Centocor Inc., USA

World Scientific

NEW JERSEY · LONDON · SINGAPORE · BEIJING · SHANGHAI · HONG KONG · TAIPEI · CHENNAI

Published by

World Scientific Publishing Co. Pte. Ltd.

5 Toh Tuck Link, Singapore 596224

USA office: 27 Warren Street, Suite 401-402, Hackensack, NJ 07601

UK office: 57 Shelton Street, Covent Garden, London WC2H 9HE

Library of Congress Cataloging-in-Publication Data
Raghavarao, Damaraju.
    Block designs : analysis, combinatorics, and applications / Damaraju Raghavarao,
Lakshmi V. Padgett.
       p. cm. -- (Series on applied mathematics ; v. 17)
    Includes bibliographical references and indexes.
    ISBN 981-256-360-1 (alk. paper)
    1. Block designs. I. Padgett, Lakshmi V. II. Title. III. Series.

QA279.R337 2005
519.5'7--dc22

                             2005050601

British Library Cataloguing-in-Publication Data
A catalogue record for this book is available from the British Library.

Typeset by Stallion Press
Email: enquiries@stallionpress.com

Printed in Singapore by World Scientific Printers (S) Pte Ltd

To Sharada, Venkatrayudu, and Darryl

DR and LP

Contents

Preface

Design of Experiments is a fascinating subject. Combinatorial mathematicians and statisticians made significant contributions for its development.

There are many books on the subject at different levels for different types of audiences. Raghavarao in 1971 published the monograph, "Constructions and Combinatorial Problems in Design of Experiments," discussing most of the design combinatorics useful for statisticians. Block designs and especially partially balanced incomplete block designs were well discussed in that work.

The designs originally developed with a specific purpose are useful to answer a completely different problem. There are applications of standard designs as finite sample support, to prepare questionnaires to elicit information on sensitive issues, to determine respondents evaluation of different attributes, group testing, etc. All these diversified applications are not so far documented in a single place.

The main objective of this monograph is to update the constructions and combinatorial aspects of block designs from the 1971 book, bring together the diversified applications, and elegantly provide the mathematics of the statistical analysis of block designs.

The authors expect that this will be a useful reference monograph for researchers working on experimental designs and related areas. This will also be useful as a text for a special topics graduate course. By seeing the applications discussed in this work, the serious readers may extend the ideas to micro-array experiments, group testing, behavioral experiments, designs to study drug interactions, etc.

A strong matrix background, with some exposure to mathematical statistics and basic experimental designs is required to follow this monograph completely. Mathematicians without statistics background may follow most of the material except few statistical applications. We hope that this monograph will be useful to both statisticians and combinatorial mathematicians.

Damaraju Raghavarao
Lakshmi V. Padgett

March 2005

1

Linear Estimation and Tests for Linear Hypotheses

1.1 The Model

Data, from experiments, following normal distribution are analyzed using a linear model. If $Y_1, Y_2, \ldots, Y_n$ are random variables corresponding to n observations, Y_i is modelled as a known linear combination of unknown parameters and the unexplained part is attributed to a random error. We thus assume

$$Y_i = \sum_{j=1}^{p} x_{ij}\beta_j + e_i, \qquad (1.1)$$

where x_{ij} are known constants, β_j are unknown parameters, and e_i are random errors. Let $\mathbf{Y}' = (Y_1, Y_2, \ldots, Y_n)$, $\beta' = (\beta_1, \beta_2, \ldots, \beta_p)$, $\mathbf{e}' = (e_1, e_2, \ldots, e_n)$ and $X = (x_{ij})$ be an $n \times p$ matrix. Then (1.1) can be written in matrix notation as

$$\mathbf{Y} = X\beta + \mathbf{e}. \qquad (1.2)$$

Matrix X is called the design matrix. We are interested in estimating and testing hypotheses about linear parametric functions of β. To this end we need to assume

1. $E(\mathbf{e}) = \mathbf{0}$,
2. $\mathrm{Var}(\mathbf{e}) = \sigma^2 I_n$ or $\sigma^2 V$,
3. $\mathbf{e}$ has a normal distribution,

where I_n is the $n \times n$ identity matrix, $E(\bullet)$ and $\mathrm{Var}(\bullet)$ are the expected value and the dispersion matrix of the vector in parenthesis. In assumption 2, V is a known positive definite matrix. The assumption of $\sigma^2 V$ to $\mathrm{Var}(\mathbf{e})$ can be reduced to $\sigma^2 I_n$ by transforming

$$V^{-\frac{1}{2}}\mathbf{Y} = V^{-\frac{1}{2}}X\beta + V^{-\frac{1}{2}}\mathbf{e},$$
$$\mathbf{Y}^* = X^*\beta + \mathbf{e}^*, \qquad (1.3)$$

1

so that $\text{Var}(\mathbf{e}^*) = \sigma^2 I_n$. The assumptions 1 and 2 are needed to get point estimates for linear parametric functions of $\boldsymbol{\beta}$, whereas the assumption 3 is needed to get confidence intervals for linear parametric functions of $\boldsymbol{\beta}$ as well as testing hypotheses about linear parametric functions of $\boldsymbol{\beta}$ as needed. We write the linear model (1.2) shortly as $(\mathbf{Y}, X\boldsymbol{\beta}, \sigma^2 I_n)$, or $(\mathbf{Y}, X\boldsymbol{\beta}, \sigma^2 V)$.

When the design matrix X has only 1 or 0 values indicating the presence or absence of the parameters, the model (1.2) is called an Analysis of Variance Model. If X has quantitative values, the model is called a Regression Model. If some columns of X have indicator 1 or 0 values and some columns have quantitative values, the model is called an Analysis of Covariance Model.

The parameters β_j may be fixed effects, random effects, or a mixture of fixed and random effects.

In the sequel, we assume that the experimenter is interested in drawing inferences on the specific treatments used in the experiment and hence the treatment effects are fixed effects. The blocks containing experimental units may be fixed or random depending on the way the blocks are selected. We will consider the block effects as fixed and add results with random block effects as needed.

1.2 Estimability and Best Linear Unbiased Estimators

We define

Definition 1.1   A linear parametric function $\ell'\boldsymbol{\beta}$ is said to be *estimable* if there exists an $n \times 1$ column vector $\mathbf{a}$ such that

$$E(\mathbf{a}'\mathbf{Y}) = \ell'\boldsymbol{\beta}. \tag{1.4}$$

Then $\mathbf{a}'\mathbf{Y}$ is said to be an unbiased estimator of $\ell'\boldsymbol{\beta}$. Equation (1.4) implies that $\mathbf{a}'X\boldsymbol{\beta} = \ell'\boldsymbol{\beta}$ for every $\boldsymbol{\beta}$ and hence

$$X'\mathbf{a} = \ell. \tag{1.5}$$

The existence of the vector $\mathbf{a}$ satisfying (1.5) is necessary and sufficient for $\ell'\boldsymbol{\beta}$ to be estimable and we have

Theorem 1.1   *A necessary and sufficient condition for the estimability of $\ell'\boldsymbol{\beta}$ is*

$$\text{Rank}(X') = \text{Rank}(X'|\ell). \tag{1.6}$$

There may be several linear unbiased estimators for an estimable parametric function $\ell'\boldsymbol{\beta}$ and of them, the one with the smallest variance is called the Best Linear Unbiased Estimator (blue). We consider the linear model $(\mathbf{Y}, X\boldsymbol{\beta}, \sigma^2 I_n)$ unless otherwise specified. For an unbiased estimator $\mathbf{a}'\mathbf{Y}$ of $\ell'\boldsymbol{\beta}$, we have $\mathrm{Var}(\mathbf{a}'\mathbf{Y}) = \sigma^2\mathbf{a}'\mathbf{a}$. We thus minimize $\mathbf{a}'\mathbf{a}$ such that (1.5) is satisfied. Using Lagrange multiplier vector $\boldsymbol{\lambda}$, considering $\mathbf{a}'\mathbf{a} - 2\boldsymbol{\lambda}'(X'\mathbf{a} - \ell)$ and differentiating with respect to vectors $\mathbf{a}$ and $\boldsymbol{\lambda}$ and equating to zero, we get $\mathbf{a} = X\boldsymbol{\lambda}$, and substituting in (1.5), we get

$$X'X\boldsymbol{\lambda} = \ell. \tag{1.7}$$

The existence of the vector $\boldsymbol{\lambda}$ satisfying (1.7) is necessary and sufficient for the existence of the blue of $\ell'\boldsymbol{\beta}$ and we have:

Theorem 1.2 *A necessary and sufficient condition for the existence of the blue for $\ell'\boldsymbol{\beta}$ is*

$$\mathrm{Rank}(X'X) = \mathrm{Rank}(X'X|\ell). \tag{1.8}$$

The conditions (1.6) and (1.8) are the same. In fact (1.8) implies (1.7) for some $\boldsymbol{\lambda}$, which implies (1.5) with $\mathbf{a} = X\boldsymbol{\lambda}$ and hence (1.6). Conversely,

$$\mathrm{Rank}(X'X) = \mathrm{Rank}(X') = \mathrm{Rank}(X'|\ell) \geq \mathrm{Rank}(X'|\ell)\begin{pmatrix} X & 0 \\ 0 & 1 \end{pmatrix}$$

$$= \mathrm{Rank}(X'X|\ell) \geq \mathrm{Rank}(X'X)$$

and hence (1.8).

The solution of (1.7) depends on the g-inverse of matrices and in this monograph we need g-inverse of symmetric matrices. Let A be a symmetric $n \times n$ matrix of rank r with nonzero eigenvalues $\lambda_1, \lambda_2, \ldots, \lambda_r$ with orthonormal eigenvectors $\boldsymbol{\xi}_1, \boldsymbol{\xi}_2, \ldots, \boldsymbol{\xi}_r$. Further, let $\boldsymbol{\eta}_1, \boldsymbol{\eta}_2, \ldots, \boldsymbol{\eta}_{n-r}$ be orthonormal eigenvectors corresponding to the zero eigenvalue of A. Then the g-inverses of A, denoted by A^-, satisfying

$$AA^-A = A, \tag{1.9}$$

are

$$A^- = \sum_{i=1}^{r} \frac{1}{\lambda_i}\boldsymbol{\xi}_i\boldsymbol{\xi}_i' \tag{1.10}$$

or

$$A^- = \left(A + \sum_{j=1}^{n-r} a_j \boldsymbol{\eta}_j \boldsymbol{\eta}'_j \right)^{-1}, \tag{1.11}$$

where $a_1, a_2, \ldots, a_{n-r}$ are positive real numbers. If

$$A = \begin{pmatrix} A_{11} & A_{12} \\ A_{21} & A_{22} \end{pmatrix},$$

where rank $A_{11} = r$, and A_{11} is a $r \times r$ matrix, then

$$A^- = \begin{bmatrix} A_{11}^{-1} & 0 \\ 0 & 0 \end{bmatrix}, \tag{1.12}$$

where 0 is a zero matrix of appropriate order. We will use one of the three forms (1.10), (1.11) or (1.12) of g-inverses as needed.

The Moore–Penrose inverse, A^+, of A satisfies

$$\begin{aligned} AA^+A &= A, \quad A^+AA^+ = A^+ \\ AA^+ &\text{ is a symmetric matix, } \quad \text{and} \\ A^+A &\text{ is a symmetric matix.} \end{aligned} \tag{1.13}$$

The Moore–Penrose inverse, A^+, is unique, whereas a g-inverse, A^-, is not unique. The matrix (1.10) may be seen as the Moore–Penrose inverse of A.

The following result is very useful.

Theorem 1.3 *We have*

1. $(X'X)(X'X)^-X' = X'$,
2. $X(X'X)^-(X'X) = X$,
3. $X(X'X)^-X'$ *is idempotent and hence* $\mathrm{tr}(X(X'X)^-X') = \mathrm{Rank}(X)$,
4. $I - X(X'X)^-X'$ *is idempotent.*

Proof. We prove 1 and 2 can be similarly proved. Consider

$$\begin{aligned} (X'X(X'X)^-X' &- X')(X(X'X)^{-'}X'X - X) \\ &= X'X - X'X - X'X + X'X = 0, \end{aligned}$$

and hence 1. To prove 3,

$$(X(X'X)^-X')(X(X'X)^-X') = X(X'X)^-X',$$

and noting that trace and rank are the same for an idempotent matrix, we have

$$\mathrm{tr}(X(X'X)^-X') = \mathrm{Rank}(X(X'X)^-X') = \mathrm{Rank}(X'X(X'X)^-)$$
$$= \mathrm{Rank}(X'X) = \mathrm{Rank}(X).$$

The fourth part of the theorem can be easily verified. We will now prove

Theorem 1.4 *The blue of $\ell'\beta$ is unique.*

Proof. Let W_1 and W_2 be any 2, g-inverses of $X'X$ and let $\lambda_1 = W_1\ell$, $\lambda_2 = W_2\ell$. Further let $\mathbf{a}_i = X\lambda_i$ for $i = 1, 2$. Then $\mathbf{a}_1'\mathbf{Y} = \mathbf{a}_2'\mathbf{Y}$ since

$$(\mathbf{a}_1 - \mathbf{a}_2)'(\mathbf{a}_1 - \mathbf{a}_2) = (\lambda_1 - \lambda_2)'X'X(\lambda_1 - \lambda_2)$$
$$= (\lambda_1' - \lambda_2')(\ell - \ell) = 0.$$

We denote the blue of $\ell'\beta$ by $\widehat{\ell'\beta}$ and we showed that $\widehat{\ell'\beta} = \lambda'X'\mathbf{Y}$ where $\lambda = (X'X)^-\ell$ for any g-inverse of $X'X$. Clearly,

$$\mathrm{Var}(\widehat{\ell'\beta}) = \mathrm{Var}(\lambda'X'Y) = \sigma^2\lambda'X'X\lambda = \sigma^2\ell'(X'X)^-\ell. \tag{1.14}$$

If $\ell'\beta$ and $\mathbf{m}'\beta$ are estimable, then $\mathrm{Cov}(\widehat{\ell'\beta}, \widehat{\mathbf{m}'\beta}) = \mathrm{Cov}(\lambda'X'Y, \mu'X'Y)$, where $\lambda = (X'X)^-\ell$ and $\mu = (X'X)^-\mathbf{m}$ and we get

$$\mathrm{Cov}(\widehat{\ell'\beta}, \widehat{\mathbf{m}'\beta}) = \sigma^2\ell'(X'X)^-\mathbf{m}. \tag{1.15}$$

Suppose L is a $p \times k$ matrix of rank k such that the k components of the vector $L'\beta$ are estimable. Then the unbiased estimators of the k components of $L'\beta$ are the components of $A'\mathbf{Y}$, where $X'A = L$, and the blues of the components of $L'\beta$ are $\Lambda'X'\mathbf{Y}$, where Λ is a $p \times k$ matrix, $\Lambda = (X'X)^-L$. Now

$$\mathrm{Var}(A'\mathbf{Y}) = \sigma^2 A'A,$$
$$\mathrm{Var}(\Lambda'X'\mathbf{Y}) = \sigma^2 L'(X'X)^-L = \sigma^2 A'X(X'X)^-X'A.$$

Since $I - X(X'X)^-X'$ is a symmetric idempotent matrix, it is at least positive semi-definite and hence

$$|A'A| \geq |A'X(X'X)^-X'A|.$$

We thus proved

Theorem 1.5 *The generalized variance of the blues of independent estimable linear parametric functions is not greater than the generalized variance of the unbiased estimators of those independent estimable linear parametric functions.*

1.3 Least Squares Estimates

The value of $\boldsymbol{\beta}$ in Model (1.2) obtained by minimizing $\mathbf{e}'\mathbf{e}$ with respect to $\boldsymbol{\beta}$ is called the Least Squares Estimator. The equations giving critical values of $\boldsymbol{\beta}$ are called normal equations and are obtained by differentiating $\mathbf{e}'\mathbf{e}$ with respect to $\boldsymbol{\beta}$ and equating to zero. We have

$$\mathbf{e}'\mathbf{e} = \mathbf{Y}'\mathbf{Y} - 2\mathbf{Y}'X\boldsymbol{\beta} + \boldsymbol{\beta}'X'X\boldsymbol{\beta}$$

and the normal equations are

$$X'X\hat{\boldsymbol{\beta}} = X'\mathbf{Y}, \tag{1.16}$$

where $\hat{\boldsymbol{\beta}}$ is the least squares estimator of $\boldsymbol{\beta}$. The minimum of $\mathbf{e}'\mathbf{e}$ is denoted by R_0^2 and it is called the unconditional minimum residual sum of squares. We have

$$R_0^2 = \min_{\boldsymbol{\beta}} \mathbf{e}'\mathbf{e} = (\mathbf{Y} - X\hat{\boldsymbol{\beta}})'(\mathbf{Y} - X\hat{\boldsymbol{\beta}}) = \mathbf{Y}'\mathbf{Y} - \hat{\boldsymbol{\beta}}'X'\mathbf{Y}. \tag{1.17}$$

Taking $\hat{\boldsymbol{\beta}}$ as $(X'X)^- X'\mathbf{Y}$ as a solution of (1.16), we can express R_0^2 as

$$R_0^2 = \mathbf{Y}'\{I_n - X(X'X)^- X'\}\mathbf{Y}.$$

We prove

Theorem 1.6 *If $\ell'\boldsymbol{\beta}$ is an estimable function, its blue $\widehat{\ell'\boldsymbol{\beta}}$ is $\ell'\hat{\boldsymbol{\beta}}$, where $\hat{\boldsymbol{\beta}}$ is any solution of (1.16). This solution is invariant to the choice of g-inverse of $X'X$.*

Proof. We proved that $\widehat{\ell'\boldsymbol{\beta}} = \boldsymbol{\lambda}'X'\mathbf{Y}$, where $\boldsymbol{\lambda}$ satisfies (1.7). Substituting for $X'\mathbf{Y}$ from (1.16) and using (1.7) we get

$$\widehat{\ell'\boldsymbol{\beta}} = \boldsymbol{\lambda}'X'\mathbf{Y} = \boldsymbol{\lambda}'X'X\hat{\boldsymbol{\beta}} = \ell'\hat{\boldsymbol{\beta}}.$$

Let W_1, W_2 be any two g-inverses of $X'X$ and let $\hat{\boldsymbol{\beta}}^{(1)} = W_1 X'\mathbf{Y}$, $\hat{\boldsymbol{\beta}}^{(2)} = W_2 X'\mathbf{Y}$. Now

$$(\ell'W_1 X' - \ell'W_2 X')(X W_1'\ell - X W_2'\ell)$$
$$= \ell'W_1'\ell - \ell'W_2'\ell - \ell'W_1'\ell + \ell'W_2'\ell = 0,$$

noting that $\ell'(X'X)^- X'X = \ell$ for an estimable parameteric function $\ell'\boldsymbol{\beta}$. We also have

Theorem 1.7   1. $E(R_0^2) = (n - r)\sigma^2$, *where* $r = \text{Rank}(X)$

2. $\text{Cov}\{\ell'\hat{\boldsymbol{\beta}}, (I_n - X(X'X)^- X')\mathbf{Y}\} = 0$, *for estimable* $\ell'\boldsymbol{\beta}$.

Proof.

$$
\begin{aligned}
E(R_0^2) &= E\{\mathbf{Y}'(I_n - X(X'X)^- X')\mathbf{Y}\} \\
&= E\{(\mathbf{Y} - X\boldsymbol{\beta})'(I_n - X(X'X)^- X')(\mathbf{Y} - X\boldsymbol{\beta})\} \\
&= E\mathrm{tr}\{(\mathbf{Y} - X\boldsymbol{\beta})'(I_n - X(X'X)^- X'))(\mathbf{Y} - X\boldsymbol{\beta})\} \\
&= E\{\mathrm{tr}[(I_n - X(X'X)^- X')(\mathbf{Y} - X\boldsymbol{\beta})(\mathbf{Y} - X\boldsymbol{\beta})']\} \\
&= \mathrm{tr}[I_n - X(X'X)^- X']\sigma^2 = \sigma^2(n - \mathrm{tr}(X'X(X'X)^-) \\
&= \sigma^2(n - r).
\end{aligned}
$$

Also

$$
\begin{aligned}
&\mathrm{Cov}(\boldsymbol{\ell}'\hat{\boldsymbol{\beta}}, (I_n - X(X'X)^- X')\mathbf{Y}) \\
&= \mathrm{Cov}(\boldsymbol{\ell}'(X'X)^-(X'\mathbf{Y}), (I_n - X(X'X)^- X')\mathbf{Y}) \\
&= \mathrm{Cov}(\boldsymbol{\ell}'(X'X)^- X'(\mathbf{Y} - X\boldsymbol{\beta}), (I_n - X(X'X)^- X')(\mathbf{Y} - X\boldsymbol{\beta})) \\
&= (\boldsymbol{\ell}'(X'X)^- X')(I_n - X(X'X)^- X')\sigma^2 \\
&= \boldsymbol{\ell}'(X'X)^-(X' - X')\sigma^2 \\
&= 0.
\end{aligned}
$$

Since $\boldsymbol{\ell}'(X'X)^-\boldsymbol{\ell}\sigma^2 = \mathrm{Var}(\widehat{\boldsymbol{\ell}'\boldsymbol{\beta}}) = \mathrm{Var}(\boldsymbol{\ell}'\hat{\boldsymbol{\beta}})$, we can treat $(X'X)^-\sigma^2$ as if it is the dispersion matrix of the least squares estimator $\hat{\boldsymbol{\beta}}$, even though the components of $\boldsymbol{\beta}$ may not be estimable.

1.4 Error Functions

We have

Definition 1.2 A linear function of the observations $\mathbf{e}'\mathbf{Y}$ is called an *error function* if $E(\mathbf{e}'\mathbf{Y}) = 0$.

Clearly

Theorem 1.8 *In a linear model* $(\mathbf{Y}, X\boldsymbol{\beta}, \sigma^2 V)$ *the function* $\mathbf{e}'\mathbf{Y}$ *is an error function if and only if*

$$
X'\mathbf{e} = 0. \tag{1.18}
$$

The error functions form a vector space of dimensionality $n - r$, *where* r *is the* Rank(X) *and* X *is the* $n \times p$ *design matrix.*

Proof. Since $E(\mathbf{e}'\mathbf{Y}) = \mathbf{e}'X\boldsymbol{\beta} = 0$, is an identity in $\boldsymbol{\beta}$ we have (1.18) as a necessary and sufficient condition for $\mathbf{e}'\mathbf{Y}$ to be an error function. The system of homogeneous equations (1.18) have n unknowns and the coefficient matrix X has rank r and hence the solutions form a vector space of dimensionality $n - r$.

Let $\mathbf{e}_i$ for $i = 1, 2, \ldots, n - r$ be orthonormal solutions of (1.18). Since

$$(I_n - X(X'X)^- X')\mathbf{e}_i = \mathbf{e}_i, \quad i = 1, 2, \ldots, n - r,$$

and $I_n - X(X'X)^- X'$ is an idempotent matrix of rank $n - r$, $\mathbf{e}_i$ are orthonormal eigenvectors corresponding to the eigenvalue 1 with multiplicity $n - r$ of $I_n - X(X'X)^- X'$, and

$$I_n - X(X'X)^- X' = \sum_{i=1}^{n-r} \mathbf{e}_i \mathbf{e}_i'. \tag{1.19}$$

Hence the minimum residual sum of squares

$$R_0^2 = \mathbf{Y}'(I - X(X'X)^- X')\mathbf{Y} = \mathbf{Y}' \left(\sum_{i=1}^{n-r} \mathbf{e}_i \mathbf{e}_i' \right) \mathbf{Y} = \sum_{i=1}^{n-r} (\mathbf{e}_i' \mathbf{Y})^2,$$

is the sum of squares of $n - r$ orthonormal error functions. Let $\mathbf{f}_j$ for $j = 1, 2, \ldots, r$ be r orthonormal eigenvectors corresponding to the zero eigenvalue of multiplicity r of $I_n - X(X'X)^- X'$. Then $\mathbf{e}_i' \mathbf{f}_j = 0$ for $i = 1, 2, \ldots, n - r; j = 1, 2, \ldots, r$, and

$$X(X'X)^- X' \mathbf{f}_j = \mathbf{f}_j.$$

Now $X(X'X)^- X'$ is an idempotent matrix of rank r and $\mathbf{f}_j$ for $j = 1, 2, \ldots, r$ are the orthonormal eigenvectors corresponding to the eigenvalue 1 with multiplicity r of $X(X'X)^- X'$. Hence

$$X(X'X)^- X' = \sum_{j=1}^{r} \mathbf{f}_j \mathbf{f}_j' \tag{1.20}$$

and

$$\mathbf{Y}' X(X'X)^- X' \mathbf{Y} = \sum_{j=1}^{r} (\mathbf{f}_j' \mathbf{Y})^2.$$

$\mathbf{f}_j' \mathbf{Y}$ are blues of $\mathbf{f}_j' X \boldsymbol{\beta}$ for $j = 1, 2, \ldots, r$ and thus $\mathbf{Y}' X(X'X)^- X' \mathbf{Y}$ is the sum of squares of r orthonormal blues.

Thus the sum of squares $\mathbf{Y}'\mathbf{Y}$ of the linear model $(\mathbf{Y}, X\boldsymbol{\beta}, \sigma^2 I_n)$ is partitioned into $\mathbf{Y}' X(X'X)^- X' \mathbf{Y}$, which is the sum of squares of r orthonormal blues and $\mathbf{Y}'(I - X(X'X)^- X')\mathbf{Y}$, which is the sum of squares of $n - r$ orthonormal error functions, uncorrelated with orthonormal blues, because $\text{Cov}(\mathbf{e}_i' \mathbf{Y}, \mathbf{f}_j' \mathbf{Y}) = \mathbf{e}_i' \mathbf{f}_j \sigma^2 = 0$.

1.5 Weighted Normal Equations

The linear model $(\mathbf{Y}, X\boldsymbol{\beta}, \sigma^2 V)$, where V is a known positive definite matrix is equivalent to the linear model $(V^{-1/2}\mathbf{Y}, V^{-1/2}X\boldsymbol{\beta}, \sigma^2 I_n)$. Hence $\ell'\boldsymbol{\beta}$ has blue if and only if

$$\text{Rank}(X'V^{-1}X) = \text{Rank}(X'V^{-1}X|\ell), \tag{1.21}$$

and the blue of $\ell'\boldsymbol{\beta}$ is $\ell'\tilde{\boldsymbol{\beta}}$, where $\tilde{\boldsymbol{\beta}}$ is a solution of the weighted normal equations

$$\begin{aligned} X'(V^{-1/2})'V^{-1/2}X\tilde{\boldsymbol{\beta}} &= X'(V^{-1/2})'V^{-1/2}\mathbf{Y}, \\ X'V^{-1}X\tilde{\boldsymbol{\beta}} &= X'V^{-1}\mathbf{Y}. \end{aligned} \tag{1.22}$$

It may be noted that (1.22) may be obtained by differentiating $(\mathbf{Y}-X\boldsymbol{\beta})'V^{-1}(\mathbf{Y}-X\boldsymbol{\beta})$ with respect to $\boldsymbol{\beta}$ and equating the derivative to zero.

We now turn our attention to the case where V is a known singular symmetric matrix. Let rank of V be t $(< n)$ and let $V = MM'$, be the rank factorization of V, where M is an $n \times t$ matrix of rank t. Then $M'M$ is an $t \times t$ non-singular matrix. We can easily verify the following:

Theorem 1.9   1. $V^+ = M(M'M)^{-2}M'$,

2. $M'V^+M = I_t$.

From the linear model $(\mathbf{Y}, X\boldsymbol{\beta}, \sigma^2 V)$, where V is a singular symmetric known matrix, we transform the observational vector $\mathbf{Y}$ to $\mathbf{U}$ given by

$$\mathbf{U} = (M'M)^{-1}M'\mathbf{Y}.$$

The transformed linear model is then $(\mathbf{U}, X^*\boldsymbol{\beta}, \sigma^2 I_t)$, where $X^* = (M'M)^{-1}M'X$. Now $\ell'\boldsymbol{\beta}$ has blue if and only if

$$\text{Rank}(X'V^+X) = \text{Rank}(X'V^+X|\ell), \tag{1.23}$$

by noting that $X^{*\prime}X^* = X'M(M'M)^{-2}M'X = X'V^+X$.

The blue of $\ell'\boldsymbol{\beta}$ is $\ell'\boldsymbol{\beta}^*$, where $\boldsymbol{\beta}^*$ is a solution of the weighted least squares

$$(X'V^+X)\boldsymbol{\beta}^* = X'V^+\mathbf{Y}, \tag{1.24}$$

by noting that $X^{*\prime}\mathbf{U} = X'M(M'M)^{-2}M'\mathbf{Y} = X'V^+\mathbf{Y}$.

1.6 Distributions of Quadratic Forms

In this and subsequent sections we assume that the errors in the linear model follow a normal distribution and the observational vector $\mathbf{Y}$ is n-dimensional. We have

Theorem 1.10 *For a linear model* $(\mathbf{Y}, X\boldsymbol{\beta}, \sigma^2 V)$, *where V is a known positive definite matrix.*

1. $(\mathbf{Y} - X\boldsymbol{\beta})' V^{-1}(\mathbf{Y} - X\boldsymbol{\beta})/\sigma^2 \sim \chi^2(n)$,
2. $\mathbf{Y}' V^{-1} \mathbf{Y}/\sigma^2 \sim \chi^{2\prime}(n, \Delta)$,   *where* $\Delta = \boldsymbol{\beta}' X' V^{-1} X\boldsymbol{\beta}/\sigma^2$,

$\chi^2(n)$ *and* $\chi^{2\prime}(n, \Delta)$ *being central chi-square distribution with n degrees of freedom and non-central chi-square distribution with n degrees of freedom and non-centrality parameter Δ, respectively.*

Proof. $\frac{1}{\sigma} V^{-1/2}(\mathbf{Y} - X\boldsymbol{\beta})$ has n-dimensional multivariate normal distribution with mean vector $\mathbf{0}$ and dispersion matrix I_n. Hence the squared length of the vector is distributed as $\chi^2(n)$. Also $\frac{1}{\sigma} V^{-1/2}\mathbf{Y} \sim N_n(\frac{1}{\sigma} V^{-1/2} X\boldsymbol{\beta}, I_n)$, where $N_n(\boldsymbol{\mu}, \Sigma)$ denotes an n-variate normal distribution with mean vector $\boldsymbol{\mu}$ and dispersion matrix Σ and hence

$$\mathbf{Y}' V^{-1} \mathbf{Y}/\sigma^2 \sim \chi^{2\prime}(n, \Delta),$$

where

$$\Delta = \left(\frac{1}{\sigma}\boldsymbol{\beta}' X' V^{-1/2}\right)\left(\frac{1}{\sigma} V^{-1/2} X\boldsymbol{\beta}\right) = \boldsymbol{\beta}' X' V^{-1} X\boldsymbol{\beta}/\sigma^2.$$

Theorem 1.11 *For the linear model* $(\mathbf{Y}, X\boldsymbol{\beta}, \sigma^2 V)$, *where V is a known positive semi definite matrix of rank $t\,(< n)$, we have*

1. $(\mathbf{Y} - X\boldsymbol{\beta})' V^{-}(\mathbf{Y} - X\boldsymbol{\beta})/\sigma^2 \sim \chi^2(t)$,
2. $\mathbf{Y}' V^{-} \mathbf{Y}/\sigma^2 \sim \chi^{2\prime}(t, \Delta)$,   *where* $\Delta = \boldsymbol{\beta}' X' V^{-} X\boldsymbol{\beta}/\sigma^2$,

V^{-} *being a symmetric g-inverse of V.*

Proof.   Let $V = MM'$ be the rank factorization of V, where M is an $n \times t$ matrix of rank t. We first prove the result using V^{+} for V^{-}. Put $\mathbf{U} = (M'M)^{-1}M'\mathbf{Y}/\sigma$, so that $\mathbf{U} \sim N_t\left(\frac{1}{\sigma}(M'M)^{-1}M'X\boldsymbol{\beta}, I_t\right)$.

Hence

$$\begin{aligned}
(\mathbf{U} - (M'M)^{-1}&M'X\boldsymbol{\beta}/\sigma)'(\mathbf{U} - (M'M)^{-1}M'X\boldsymbol{\beta}/\sigma) \\
&= (\mathbf{Y} - X\boldsymbol{\beta})' M(M'M)^{-1}(M'M)^{-1}M'(\mathbf{Y} - X\boldsymbol{\beta})/\sigma^2 \\
&= (\mathbf{Y} - X\boldsymbol{\beta})' V^{+}(\mathbf{Y} - X\boldsymbol{\beta})/\sigma^2 \sim \chi^2(t),
\end{aligned}$$

using 1 of Theorem 1.10.

For any symmetric g-inverse V^- of V, we have

$$(\mathbf{Y} - X\boldsymbol{\beta})'V^-(\mathbf{Y} - X\boldsymbol{\beta})/\sigma^2$$
$$= (\mathbf{Y} - X\boldsymbol{\beta})'V^+(\mathbf{Y} - X\boldsymbol{\beta})/\sigma^2 + (\mathbf{Y} - X\boldsymbol{\beta})'(V^- - V^+)(\mathbf{Y} - X\boldsymbol{\beta})/\sigma^2.$$
$$(1.25)$$

The mean and variance of the second quadratic form on the right-hand side are,

$$\text{Mean} = \text{tr}\{(V^- - V^+)V\} = 0,$$
$$\text{Variance} = 2\text{tr}\{(V^- - V^+)V(V^- - V^+)V\} = 0.$$

Hence the left-hand side quadratic form of (1.25) is distributed as the first quadratic form of right-hand side, that is, $\chi^2(t)$.

Part 2 of the theorem can be similarly proved.

Theorem 1.12 *If* $(\mathbf{Y}, X\boldsymbol{\beta}, \sigma^2 V)$ *is a linear model, where V is a known positive definite matrix, then* $(\mathbf{Y} - X\boldsymbol{\beta})'A(\mathbf{Y} - X\boldsymbol{\beta})/\sigma^2 \sim \chi^2(v)$, *where A is a symmetric matrix of rank v, if and only if*

$$AVAV = AV. \tag{1.26}$$

When $V = I_n$, the condition (1.26) is that A is an idempotent matrix.

Proof. Let us assume (1.26) and put $\mathbf{U} = A(\mathbf{Y} - X\boldsymbol{\beta})/\sigma$. Then $(\mathbf{Y} - X\boldsymbol{\beta})'A(\mathbf{Y} - X\boldsymbol{\beta})/\sigma^2 = (\mathbf{Y} - X\boldsymbol{\beta})'AA^-A(\mathbf{Y} - X\boldsymbol{\beta})/\sigma^2 = \mathbf{U}'A^-\mathbf{U}$, where $\mathbf{U} \sim N_n(0, A)$. From 1 of Theorem 1.11, it follows that $(\mathbf{Y} - X\boldsymbol{\beta})'A(\mathbf{Y} - X\boldsymbol{\beta})/\sigma^2 \sim \chi^2(v)$.

Conversely, assume that $(\mathbf{Y} - X\boldsymbol{\beta})'A(\mathbf{Y} - X\boldsymbol{\beta})/\sigma^2$ has a $\chi^2(v)$ distribution. Put $\mathbf{W} = V^{-1/2}(\mathbf{Y} - X\boldsymbol{\beta})/\sigma$. Then $\mathbf{W}$ is distributed $N_n(0, I_n)$. Now $(\mathbf{Y} - X\boldsymbol{\beta})'A(\mathbf{Y} - X\boldsymbol{\beta})/\sigma^2 = \mathbf{W}'V^{1/2}AV^{1/2}\mathbf{W}$, is distributed as $\chi^2(v)$. Since A is of rank v, let $\lambda_1, \lambda_2, \ldots, \lambda_v$ be the nonzero eigenvalues of $V^{1/2}AV^{1/2}$ and P be an orthogonal matrix such that

$$P'V^{1/2}AV^{1/2}P = D(\lambda_1, \lambda_2, \ldots, \lambda_v, 0, 0, \ldots, 0), \tag{1.27}$$

where $D(\bullet, \bullet, \ldots, \bullet)$ is a diagonal matrix of its arguments. Put $\mathbf{W}^* = P'\mathbf{W}$. Then $\mathbf{W}^*$ is distributed as $N_n(0, I_n)$, and

$$\mathbf{W}'V^{1/2}AV^{1/2}\mathbf{W} = \mathbf{W}^{*'}D(\lambda_1, \lambda_2, \ldots, \lambda_v, 0, 0, \ldots, 0)\mathbf{W}^* = \sum_{i=0}^{v}\lambda_i W_i^{*2},$$

where W_i^* is the ith component of $\mathbf{W}^*$. The moment generating function (mgf) of $\sum_{i=1}^{\nu} \lambda_i W_i^{*2}$, noting that W_i^* are independently distributed $N(0, 1)$ variables, is

$$M_{\sum_{i=1}^{\nu} \lambda_i W_i^{*2}}(t) = \frac{1}{\prod_{i=1}^{\nu}(1 - 2\lambda_i t)^{1/2}}$$

and this is identically the mgf of $\chi^2(\nu)$ variable which is $1/(1 - 2t)^{\nu/2}$. Hence $\lambda_i = 1$, for $i = 1, 2, \ldots, \nu$ and $V^{1/2} A V^{1/2}$ is an idempotent matrix so that

$$V^{1/2} A V^{1/2} V^{1/2} A V^{1/2} = V^{1/2} A V^{1/2},$$

which is the same as Eq. (1.26).

When $V = I_n$, Eq. (1.26) clearly reduces to the fact that A is an idempotent matrix.

Finally

Theorem 1.13 *If* $(\mathbf{Y}, X\boldsymbol{\beta}, \sigma^2 V)$ *is a linear model, where V is a known positive definite matrix and if* $(\mathbf{Y} - X\boldsymbol{\beta})' A_i (\mathbf{Y} - X\boldsymbol{\beta})/\sigma^2 \sim \chi^2(\nu_i)$ *where ν_i is the rank of A_i, for $i = 1, 2$, then both chi-square variables are independently distributed if and only if*

$$A_1 V A_2 = 0. \tag{1.28}$$

If $V = I_n$, condition (1.28) reduces to $A_1 A_2 = 0$.

Proof. Since $(\mathbf{Y} - X\boldsymbol{\beta})' A_i (\mathbf{Y} - X\boldsymbol{\beta})/\sigma^2 \sim \chi^2(\nu_i)$ for $i = 1, 2$, we have

$$A_i V A_i = A_i, \quad i = 1, 2 \tag{1.29}$$

and if these two chi-square variables are independent, their sum is a χ^2 variable which implies that

$$(A_1 + A_2)V(A_1 + A_2) = A_1 + A_2. \tag{1.30}$$

Condition (1.30) in the light of (1.29) implies that

$$A_1 V A_2 = -A_2 V A_1. \tag{1.31}$$

Pre-multiplying (1.31) by $A_1 V$, we get

$$A_1 V A_2 = -A_1 V A_2 V A_1 \tag{1.32}$$

and post-multiplying (1.31) by $V A_1$, we get

$$A_2 V A_1 = -A_1 V A_2 V A_1. \tag{1.33}$$

Equations (1.32) and (1.33) imply that $A_1 V A_2 = A_2 V A_1$ and this result in conjunction with (1.31) implies (1.28).

Conversely assume that $(\mathbf{Y} - X\boldsymbol{\beta})'A_i(\mathbf{Y} - X\boldsymbol{\beta})/\sigma^2 \sim \chi^2(\nu_i)$ for $i = 1, 2$ and (1.28) is true. Then $(A_1 + A_2)V(A_1 + A_2) = (A_1 + A_2)$ and consequently $(\mathbf{Y} - X\boldsymbol{\beta})'(A_1 + A_2)(\mathbf{Y} - X\boldsymbol{\beta})/\sigma^2 \sim \chi^2(\nu_1 + \nu_2)$ where $\nu_1 + \nu_2 = \text{Rank}(A_1 + A_2) = \text{trace}(A_1 + A_2)$. Since the mgf of $(\mathbf{Y} - X\boldsymbol{\beta})'(A_1 + A_2)(\mathbf{Y} - X\boldsymbol{\beta})/\sigma^2$ is the product of mgfs of $(\mathbf{Y} - X\boldsymbol{\beta})'(A_i)(\mathbf{Y} - X\boldsymbol{\beta})/\sigma^2$ for $i = 1, 2$, we have that $(\mathbf{Y} - X\boldsymbol{\beta})'A_1(\mathbf{Y} - X\boldsymbol{\beta})/\sigma^2$ is independently distributed of $(\mathbf{Y} - X\boldsymbol{\beta})'A_2(\mathbf{Y} - X\boldsymbol{\beta})/\sigma^2$.

1.7 Tests of Linear Hypotheses

The minimum residual sum of squares for the linear model $(\mathbf{Y}, X\boldsymbol{\beta}, \sigma^2 I_n)$ is given by

$$R_0^2 = \mathbf{Y}'(I - X(X'X)^- X')\mathbf{Y}$$
$$= (\mathbf{Y} - X\boldsymbol{\beta})'(I - X(X'X)^- X')(\mathbf{Y} - X\boldsymbol{\beta}),$$

and as $I - X(X'X)^- X'$ is an idempotent matrix of rank $n - r$, where r is the rank of X, in view of Theorem 1.12 we have $R_0^2/\sigma^2 \sim \chi^2(n - r)$.

Let L be a $p \times k$ matrix of rank $k(\leq p)$ and the k components of $L'\boldsymbol{\beta}$ are estimable. Then the blue of $L'\boldsymbol{\beta}$ is $L'\hat{\boldsymbol{\beta}}$, where $\hat{\boldsymbol{\beta}}$ is a solution of the normal equations and further $L'\hat{\boldsymbol{\beta}} \sim N_k(L'\boldsymbol{\beta}, L'(X'X)^- L\sigma^2)$. It can be verified that $L'(X'X)^- L$ is non-singular. Wald's test statistic for testing

$$H_0 : L'\boldsymbol{\beta} = \boldsymbol{\ell}_0, \; H_A : L'\boldsymbol{\beta} \neq \boldsymbol{\ell}_0 \tag{1.34}$$

is $(L'\hat{\boldsymbol{\beta}} - \boldsymbol{\ell}_0)'(L'(X'X)^- L)^{-1}(L'\hat{\boldsymbol{\beta}} - \boldsymbol{\ell}_0)/\sigma^2$ and this has $\chi^2(k)$ distribution under H_0 of (1.34). Since Wald's test statistic has the nuisance parameter σ^2, we use the statistic

$$F = \frac{(L'\hat{\boldsymbol{\beta}} - \boldsymbol{\ell}_0)'\{L'(X'X)^- L\}^{-1}(L'\hat{\boldsymbol{\beta}} - \boldsymbol{\ell}_0)/k}{R_0^2/(n - r)}, \tag{1.35}$$

which has F distribution with k and $n - r$ degrees of freedom, as the numerator and denominator variables of (1.35) are independent because $\{I - X(X'X)^- X'\}X = 0$. Here $(L'\hat{\boldsymbol{\beta}} - \boldsymbol{\ell}_0)'(L'(X'X)^- L)^{-1}(L'\hat{\boldsymbol{\beta}} - \boldsymbol{\ell}_0)$ is called the sum of squares of H_0, denoted by SS_{H_0}. The hypotheses degrees of freedom is k. We call SS_{H_0}/k as the Mean Square of H_0 denoted by MS_{H_0}. R_0^2 is called the Error Sum of Squares denoted by SS_e and SS_e is based on $n - r$ degrees of freedom. The Error Mean Square, $MS_e = R_0^2/(n - r)$.

We thus have

Theorem 1.14　*The critical region for testing the null hypothesis of* (1.34) *is*

$$F > F_{1-\alpha}(k, n - r), \tag{1.36}$$

where F is given by (1.35) *and* $F_{1-\alpha}(k, n - r)$ *is the* $(1 - \alpha)100$ *percentile point of an F distribution with k numerator and* $n - r$ *denominator degrees of freedom.*

Alternatively, the p-value for testing the hypotheses (1.34) is

$$p\text{-value} = P(F(k, n - r) \geq F_{\text{cal}}), \tag{1.37}$$

where $F(k, n - r)$ is the F-variable with k and $n - r$ degrees of freedom, F_{cal} is the calculated F statistic of (1.35) and $P(\bullet)$ is the probability of the statement in the parentheses.

We now give an alternative convenient expression for SS_{H_0}. Let R_1^2 be the minimum of $\mathbf{e}'\mathbf{e}$ under H_0. Here

$$R_1^2 = \min_{\substack{\boldsymbol{\beta} \\ L'\boldsymbol{\beta} = \ell_0}} (\mathbf{Y} - X\boldsymbol{\beta})'(\mathbf{Y} - X\boldsymbol{\beta}).$$

Let the minimum for R_1^2 occur when $\boldsymbol{\beta} = \tilde{\boldsymbol{\beta}}$. Then $\tilde{\boldsymbol{\beta}}$ satisfies the equation

$$(X'X)\tilde{\boldsymbol{\beta}} = X'\mathbf{Y} + L\omega, \tag{1.38}$$

where ω is a vector of Lagrange multipliers. Using the normal equations (1.16), we can rewrite (1.38) as

$$(X'X)(\tilde{\boldsymbol{\beta}} - \hat{\boldsymbol{\beta}}) = L\omega. \tag{1.39}$$

Solving for ω in (1.39), we get

$$\omega = (L'(X'X)^- L)^{-1} L'(\tilde{\boldsymbol{\beta}} - \hat{\boldsymbol{\beta}}) \tag{1.40}$$

noting that $L'(X'X)^- L$ is non-singular.

Hence

$$
\begin{aligned}
R_1^2 &= (\mathbf{Y} - X\tilde{\boldsymbol{\beta}})'(\mathbf{Y} - X\tilde{\boldsymbol{\beta}}) \\
&= (\mathbf{Y} - X\hat{\boldsymbol{\beta}} + X(\hat{\boldsymbol{\beta}} - \tilde{\boldsymbol{\beta}}))'(\mathbf{Y} - X\hat{\boldsymbol{\beta}} + X(\hat{\boldsymbol{\beta}} - \tilde{\boldsymbol{\beta}})) \\
&= R_0^2 + (\hat{\boldsymbol{\beta}} - \tilde{\boldsymbol{\beta}})'X'X(\hat{\boldsymbol{\beta}} - \tilde{\boldsymbol{\beta}}) \\
&= R_0^2 + (\hat{\boldsymbol{\beta}} - \tilde{\boldsymbol{\beta}})'L(L'(X'X)^- L)^{-1}(L'\hat{\boldsymbol{\beta}} - L'\tilde{\boldsymbol{\beta}}) \\
&= R_0^2 + (L'\hat{\boldsymbol{\beta}} - \ell_0)'(L'(X'X)^- L)^{-1}(L'\hat{\boldsymbol{\beta}} - \ell_0) \\
&= R_0^2 + SS_{H_0}.
\end{aligned}
\tag{1.41}
$$

Thus $SS_{H_0} = R_1^2 - R_0^2$. Here R_1^2 is the minimum residual sum of squares conditional on H_0, whereas R_0^2 is the unconditional minimum residual sum of squares.

Let us consider the power of the test in testing the hypotheses (1.34). Let $L'\boldsymbol{\beta} = \ell_1 (\neq \ell_0)$. Then

$$
(\widehat{L'\boldsymbol{\beta}} - \ell_0)'(L'(X'X)^- L)^{-1}(\widehat{L'\boldsymbol{\beta}} - \ell_0)/\sigma^2 \sim \chi^{2'}(k, \Delta),
\tag{1.42}
$$

where

$$
\Delta = (\ell_1 - \ell_0)'(L'(X'X)^- L)^{-1}(\ell_1 - \ell_0)'/\sigma^2.
\tag{1.43}
$$

Consequently the test statistic F has a non-central F distribution, $F'(k, n-r, \Delta)$, with k numerator and $n - r$ denominator degrees of freedom and non-centrality parameter Δ. Thus the power of the F-test, (1.35), when $L'\boldsymbol{\beta} = \ell_1 (\neq \ell_0)$ is

$$
\text{Power} = P(F'(k, n - r, \Delta) \geq F_{1-\alpha}(k, n - r)).
\tag{1.44}
$$

For further results, the interested reader is referred to C. R. Rao (1973).

2

General Analysis of Block Designs

2.1 Preliminaries

Let n experimental units be divided into b blocks of sizes $k_1, k_2, \ldots, k_b$ where the blocks consist of homogeneous units for extraneous variability. Let v treatments be applied to the units so that each unit receives one treatment. Let the ith treatment be applied to r_i experimental units. Let $\mathbf{k}' = (k_1, k_2, \ldots, k_b)$ and $\mathbf{r}' = (r_1, r_2, \ldots, r_v)$. Then $\mathbf{k}'\mathbf{1}_b = \mathbf{r}'\mathbf{1}_v = n$, where $\mathbf{1}_m$ is a column vector of 1's of dimensionality m. The treatments assigned to a block are randomly allocated to the units of that block.

Let n_{ij} be the number of units in the jth block receiving the ith treatment. The incidence relationship of the treatments in blocks can be shown by the $v \times b$ incidence matrix $N = (n_{ij})$. Then $N\mathbf{1}_b = \mathbf{r}$ and $\mathbf{1}'_v N = \mathbf{k}'$. If $n_{ij} = 1$ for every i and j, the block design is known as a Randomized Block Design. If $n_{ij} = 0$ for some (i, j), the design is called incomplete block design and there are several classes of useful incomplete block designs available in the literature.

The analysis of block designs, when the block effects are fixed is called intra-block analysis. In incomplete block designs, when the block effects are random, the block total responses also provide some information on the treatment effects and incorporating that information in the intrablock analysis we get the analysis with recovery of interblock information.

Let k_i units of the ith block be numbered from 1 to k_i. Let Y_{ij} be the response from the jth unit of the ith block and let $d(i, j)$ be the treatment applied to the jth unit of the ith block, $j = 1, 2, \ldots, k_i; i = 1, 2, \ldots, b$. We assume the linear model

$$Y_{ij} = \mu + \beta_i + \tau_{d(i,j)} + e_{ij}, \tag{2.1}$$

where μ is the general mean, β_i is the ith fixed block effect, $\tau_{d(i,j)}$ is the fixed treatment effect of the treatment $d(i, j)$ and e_{ij} are random errors assumed to be independently and normally distributed with mean zero and unknown variance σ^2. Let $\mathbf{Y}' = (Y_{11}, Y_{12}, \ldots, Y_{1k_1}, Y_{21}, Y_{22}, \ldots, Y_{2k_2}, \ldots, Y_{b1}, Y_{b2}, \ldots, Y_{bk_b})$.

Define an $n \times v$ matrix U where the rows are labelled by ij in lexicographic order for $j = 1, 2, \ldots, k_i; i = 1, 2, \ldots, b$ with the entry in the (ij)th row and ℓth column as 1 if the ℓth treatment occurs in the jth unit of the ith block, and 0, otherwise. Let the error vector $\mathbf{e}$ be defined similar to $\mathbf{Y}$. The entire observational setup, of (2.1) can be written as

$$\mathbf{Y} = [\mathbf{1}_n | D(\mathbf{1}_{k_1}, \mathbf{1}_{k_2}, \ldots, \mathbf{1}_{k_b}) | U] \begin{bmatrix} \mu \\ \beta \\ \tau \end{bmatrix} + \mathbf{e}, \tag{2.2}$$

where $\boldsymbol{\beta}' = (\beta_1, \beta_2, \ldots, \beta_b)$, $\boldsymbol{\tau}' = (\tau_1, \tau_2, \ldots, \tau_v)$ and $D(\bullet, \bullet, \ldots, \bullet)$ is a diagonal matrix of the arguments.

2.2 Intrablock Analysis of Connected Designs

For the model (2.2), the normal equations estimating $\mu, \boldsymbol{\beta}, \boldsymbol{\tau}$ are

$$\begin{bmatrix} n & \mathbf{k}' & \mathbf{r}' \\ \mathbf{k} & D(k_1, k_2, \ldots, k_b) & N' \\ \mathbf{r} & N & D(r_1, r_2, \ldots, r_v) \end{bmatrix} \begin{bmatrix} \hat{\mu} \\ \hat{\boldsymbol{\beta}} \\ \hat{\boldsymbol{\tau}} \end{bmatrix} = \begin{bmatrix} G \\ \mathbf{B} \\ \mathbf{T} \end{bmatrix}, \tag{2.3}$$

where $\mathbf{B}' = (B_1, B_2, \ldots, B_b)$, $\mathbf{T}' = (T_1, T_2, \ldots, T_v)$, $G = \sum_{i,j} Y_{ij}$, $B_i = \sum_j Y_{ij}$, $T_\ell = \sum_{\substack{i,j \\ d(i,j)=\ell}} Y_{ij}$.

Note that $\mathbf{B}$ and $\mathbf{T}$ are the column vectors of the block total responses and treatment total responses.

Solving Eq. (2.3) with the help of a g-inverse is the same as solving them iteratively. From the second component of Eq. (2.3), we have

$$\hat{\boldsymbol{\beta}} = D\left(\frac{1}{k_1}, \frac{1}{k_2}, \ldots, \frac{1}{k_b}\right) \left[\mathbf{B} - \mathbf{k}\hat{\mu} - N'\hat{\boldsymbol{\tau}}\right]$$

and substituting this in the third component, we get

$$\mathbf{r}\hat{\mu} + ND\left(\frac{1}{k_1}, \frac{1}{k_2}, \ldots, \frac{1}{k_b}\right) \left[\mathbf{B} - \mathbf{k}\hat{\mu} - N'\hat{\boldsymbol{\tau}}\right] + D(r_1, r_2, \ldots, r_v)\hat{\boldsymbol{\tau}} = \mathbf{T},$$

which simplifies to

$$C_{\tau|\beta}\hat{\boldsymbol{\tau}} = \mathbf{Q}_{\tau|\beta}, \tag{2.4}$$

where

$$C_{\tau|\beta} = D(r_1, r_2, \ldots, r_v) - ND\left(\frac{1}{k_1}, \frac{1}{k_2}, \ldots, \frac{1}{k_b}\right)N', \tag{2.5}$$

$$\mathbf{Q}_{\tau|\beta} = \mathbf{T} - ND\left(\frac{1}{k_1}, \frac{1}{k_2}, \ldots, \frac{1}{k_b}\right)\mathbf{B}.$$

$C_{\tau|\beta}$ is called the information matrix (C-matrix) for estimating the treatment effects eliminating the block effects and $\mathbf{Q}_{\tau|\beta}$ is called the adjusted treatment totals vector eliminating the block effects. The ℓth component of $\mathbf{Q}_{\tau|\beta}$ is the sum of the responses measured from the block means, for the units receiving the ℓth treatment. It can easily be verified that

$$E(\mathbf{Q}_{\tau|\beta}) = C_{\tau|\beta}\tau.$$

Equations (2.3) can also be solved for $\hat{\tau}$ first and then $\hat{\beta}$ to get

$$C_{\beta|\tau}\hat{\beta} = \mathbf{Q}_{\beta|\tau} \tag{2.6}$$

where

$$
\begin{aligned}
C_{\beta|\tau} &= D(k_1, k_2, \ldots, k_b) - N'D\left(\frac{1}{r_1}, \frac{1}{r_2}, \ldots, \frac{1}{r_v}\right)N, \\
\mathbf{Q}_{\beta|\tau} &= \mathbf{B} - N'D\left(\frac{1}{r_1}, \frac{1}{r_2}, \ldots, \frac{1}{r_v}\right)\mathbf{T}.
\end{aligned}
\tag{2.7}
$$

Now

$$
\begin{aligned}
R_0^2 &= \sum_{i,j} Y_{ij}^2 - \hat{\mu}G - \hat{\beta}'\mathbf{B} - \hat{\tau}'\mathbf{T} \\
&= \sum_{i,j} Y_{ij}^2 - \hat{\mu}G - [\mathbf{B} - k\hat{\mu} - N'\hat{\tau}]'D\left(\frac{1}{k_1}, \frac{1}{k_2}, \ldots, \frac{1}{k_b}\right)\mathbf{B} - \hat{\tau}'\mathbf{T} \\
&= \sum_{i,j} Y_{ij}^2 - \sum_i \frac{B_i^2}{k_i} - \mathbf{Q}'_{\tau|\beta}\hat{\tau} \\
&= \left(\sum_{i,j} Y_{ij}^2 - \frac{G^2}{n}\right) - \left(\sum_i \frac{B_i^2}{k_i} - \frac{G^2}{n}\right) - \mathbf{Q}'_{\tau|\beta}\hat{\tau}.
\end{aligned}
\tag{2.8}
$$

In Eq. (2.8), $\sum_{i,j} Y_{ij}^2 - G^2/n$ is known as the total sum of squares adjusted for the mean, shortly, SS_T. $\sum_i \frac{B_i^2}{k_i} - \frac{G^2}{n}$ is known as the block sum of squares adjusted for the mean and ignoring treatments, shortly, SS_B, and $\mathbf{Q}'_{\tau|\beta}\hat{\tau}$ is the treatment sum of squares adjusted for the mean and blocks, written $SS_{Tr|B}$. Analogous to (2.8), we can also get

$$R_0^2 = SS_T - SS_{Tr} - SS_{B|Tr}, \tag{2.9}$$

where $SS_{Tr} = \sum_\ell \frac{T_\ell^2}{r_\ell} - \frac{G^2}{n}$ and $SS_{B|Tr} = \mathbf{Q}'_{\beta|\tau}\hat{\beta}$.

In SAS package, using PROC GLM, if the model is written as Response = Blocks Treatments, the Type I sum of squares for blocks and treatments are respectively SS_B and $SS_{TR|B}$ and the Type III sum of squares for blocks and treatments are respectively $SS_{B|TR}$ and $SS_{Tr|B}$. If the model is written as Response = Treatments Blocks, the Type I sum of squares for treatments is SS_{TR}. The error sum of squares, SS_e is R_0^2, and the total sum of squares is SS_T.

We will primarily be interested in testing the null hypothesis of equality of treatment effects and this is possible if and only if all functions of the type $\tau_i - \tau_j (i, j = 1, 2, \ldots, v, i \neq j)$ are estimable.

We define

Definition 2.1 A linear parametric function of treatment effects, $\ell'\tau$ is called a *contrast* if $\ell'\mathbf{1}_v = 0$. It is called an elementary contrast if ℓ has only two nonzero entries consisting of 1 and -1.

Contrasts of block effects can be similarly defined.

The first part of the following theorem is proved in Raghavarao (1971) and the second part can be similarly proved.

Theorem 2.1 *All elementary contrasts of treatment effects are estimable if and only if rank of $C_{\tau|\beta}$ is $v - 1$ and all elementary contrasts of block effects are estimable if and only if rank $C_{\beta|\tau}$ is $b - 1$.*

Clearly

Theorem 2.2 *The rank of $C_{\tau|\beta}$ is $v - 1$ if and only if the rank of $C_{\beta|\tau}$ is $b - 1$.*

Proof. By sweeping out the rows in two ways, it can be shown that the rank of the coefficient matrix of the parameters in Eq. (2.3) is

$$b + \text{Rank}(C_{\tau|\beta}) = v + \text{Rank}(C_{\beta|\tau}) \tag{2.10}$$

and hence the theorem.

We define

Definition 2.2 A block design is said to be *connected* if all elementary contrasts of treatment effects as well as elementary contrasts of block effects are estimable.

For a connected design the rank of $C_{\tau|\beta}$ is $v - 1$, the rank of $C_{\beta|\tau}$ is $b - 1$ and

$$C_{\tau|\beta}\mathbf{1}_v = \mathbf{0}, \quad C_{\beta|\tau}\mathbf{1}_b = \mathbf{0}. \tag{2.11}$$

It can also be shown that for a connected design, given any two treatments θ and ϕ, there exists a chain of treatments $\theta = \theta_0, \theta_1, \theta_2, \ldots, \theta_m, \theta_{m+1} = \phi$ such that θ_i and θ_{i+1} occur together in a block for $i = 0, 1, \ldots, m$.

Eccleston and Hedayat (1974) generalized the connectedness property to globally and pseudo-globally connectedness in block designs. They showed that under certain restrictions and constraints, the class of globally connected designs contains the optimum design, which we will introduce in Sec. 2.10. In this monograph we consider only connected designs.

For a connected design, we want to test the null hypothesis

$$H_0(\text{Tr}): \tau_1 = \tau_2 = \cdots = \tau_v, \quad H_A(\text{Tr}): \tau_i \neq \tau_j \quad \text{for some } i \neq j. \tag{2.12}$$

Under the null hypothesis given in (2.12), the model (2.3) reduces to

$$\mathbf{Y} = \left[\mathbf{1}_n \,\middle|\, D(\mathbf{1}_{k_1}, \mathbf{1}_{k_2}, \ldots, \mathbf{1}_{k_b})\right] \begin{bmatrix} \mu \\ \beta \end{bmatrix} + \mathbf{e}, \tag{2.13}$$

giving rise to the normal equations

$$\begin{bmatrix} n & \mathbf{k}' \\ \mathbf{k} & D(k_1, k_2, \ldots, k_b) \end{bmatrix} \begin{bmatrix} \tilde{\mu} \\ \tilde{\beta} \end{bmatrix} = \begin{bmatrix} G \\ \mathbf{B} \end{bmatrix}, \tag{2.14}$$

with solution $\tilde{\mu} = 0$, $\tilde{\beta} = D\left(\frac{1}{k_1}, \frac{1}{k_2}, \ldots, \frac{1}{k_b}\right)\mathbf{B}$. Hence the constrained minimum residual sum of squares is

$$R_1^2 = \left(\sum_{i,j} Y_{ij}{}^2 - \frac{G^2}{n}\right) - \left(\sum_i \frac{B_i^2}{k_i} - \frac{G^2}{n}\right) \tag{2.15}$$

and

$$SS_{H_0(\text{Tr})} = R_1^2 - R_0^2 = \hat{\tau}' \mathbf{Q}_{\tau|\beta} = \mathbf{Q}'_{\tau|\beta} C^-_{\tau|\beta} \mathbf{Q}_{\tau|\beta}. \tag{2.16}$$

In $H_0(\text{Tr})$ there are $v - 1$ independent estimable elementary contrasts and hence $SS_{H_0(\text{Tr})}$ has $v - 1$ degrees of freedom. This degrees of freedom is the difference in the ranks of the coefficient matrices of the parameters vectors in Eq. (2.3) and (2.14). The unconditional minimum residual sum of squares, R_0^2, has $n - b - \text{Rank}(C_{\tau|\beta}) = n - b - v + 1$ degrees of freedom. Thus the test statistic for testing the null hypothesis of (2.12) is

$$F(\text{Tr}) = \frac{\mathbf{Q}'_{\tau|\beta} C^-_{\tau|\beta} \mathbf{Q}_{\tau|\beta}/(v - 1)}{\left[\sum_{i,j} Y_{ij}^2 - \sum_i \frac{B_i^2}{k_i} - \mathbf{Q}'_{\tau|\beta} C^-_{\tau|\beta} \mathbf{Q}_{\tau|\beta}\right] \middle/ (n - b - v + 1)}. \tag{2.17}$$

The $F(\text{Tr})$ statistic of (2.17) has an F distribution with $v - 1$ numerator and $n - b - v + 1$ denominator degrees of freedom. The p-value for testing the null hypothesis of (2.12) is

$$p_1 = p\text{-value} = P(F(v - 1, n - b - v + 1) > F_{\text{cal}}(\text{Tr})), \tag{2.18}$$

where $F_{\text{cal}}(\text{Tr})$ is the calculated $F(\text{Tr})$ statistic of (2.17) and $F(v - 1, n - b - v + 1)$ is the F variable with $v - 1$ numerator and $n - b - v + 1$ denominator degrees of freedom. The null hypothesis of (2.12) is rejected if $p_1 < \alpha$, the selected significance level.

Analogously, the equality of block effects specified by the null hypothesis

$$H_0(B): \beta_1 = \beta_2 = \cdots = \beta_b, \quad H_A(B): \beta_i \neq \beta_j \quad \text{for some } i \neq j, \tag{2.19}$$

can be tested using the statistic

$$F(B) = \frac{\mathbf{Q}'_{\beta|\tau} C^-_{\beta|\tau} \mathbf{Q}_{\beta|\tau} / (b - 1)}{\left[\sum_{i,j} Y^2_{ij} - \sum_i \frac{B^2_i}{k_i} - \mathbf{Q}'_{\tau|\beta} C^-_{\tau|\beta} \mathbf{Q}_{\tau|\beta} \right] / (n - b - v + 1)} \tag{2.20}$$

and the p-value for the test is

$$p_2 = p\text{-value} = P(F(b - 1, n - b - v + 1) > F_{\text{cal}}(B)), \tag{2.21}$$

where $F_{\text{cal}}(B)$ is the calculated $F(B)$ statistic of (2.20). The results can be summarized in Tables 2.1 and 2.2.

Table 2.1. ANOVA.

Source	df	SS	MS
Model	$b + v - 2$	$\sum_i \dfrac{B^2_i}{k_i} - \dfrac{G^2}{n} + \mathbf{Q}'_{\tau|\beta} C^-_{\tau|\beta} \mathbf{Q}_{\tau|\beta}$	
Error	$n - b - v + 1$	$R^2_0 = \sum_{i,j} Y^2_{ij} - \sum_i \dfrac{B^2_i}{k_i} - \mathbf{Q}'_{\tau|\beta} C^-_{\tau|\beta} \mathbf{Q}_{\tau|\beta}$	$R^2_0/(n - b - v + 1) = \hat{\sigma}^2$

Table 2.2. TYPE III sum of squares.

Source	df	SS	MS	F	p
Treatments	$v - 1$	$\mathbf{Q}'_{\tau|\beta} C^-_{\tau|\beta} \mathbf{Q}_{\tau|\beta}$	$\dfrac{SS_{H_0}(\text{Tr})}{v - 1} = MS_{H_0}(\text{Tr})$	$\dfrac{MS_{H_0}(\text{Tr})}{\hat{\sigma}^2}$	p_1
Blocks	$b - 1$	$\mathbf{Q}'_{\beta|\tau} C^-_{\beta|\tau} \mathbf{Q}_{\beta|\tau}$	$\dfrac{SS_{H_0}(B)}{b - 1} = MS_{H_0}(B)$	$\dfrac{MS_{H_0}(B)}{\hat{\sigma}^2}$	p_2

From the discussion in Chap. 1, $(C_{\tau|\beta} + aJ_v)^{-1}$ is a g-inverse of $C_{\tau|\beta}$, where a is a positive real number and J_v is a $v \times v$ matrix of all ones. Also, letting

$$\Omega^{-1} = C_{\tau|\beta} + \frac{1}{n}\mathbf{rr}', \tag{2.22}$$

we have $\Omega^{-1}\mathbf{1}_v = \mathbf{r}$ and hence

$$\Omega\mathbf{r} = \mathbf{1}, \tag{2.23}$$

noting that Ω^{-1} is non-singular. Hence

$$C_{\tau|\beta}\Omega C_{\tau|\beta} = \left\{\Omega^{-1} - \frac{1}{n}\mathbf{rr}'\right\}\Omega C_{\tau|\beta} = \left(I_v - \frac{1}{n}\mathbf{r1}'\right)C_{\tau|\beta} = C_{\tau|\beta}, \tag{2.24}$$

and Ω is a g-inverse of $C_{\tau|\beta}$. Using these g-inverses we have

Theorem 2.3

1. *(B.V. Shah, 1959a)* $\hat{\tau} = (C_{\tau|\beta} + aJ_v)^{-1}Q_{\tau|\beta}$
2. *(Tocher, 1952)* $\hat{\tau} = \Omega\, Q_{\tau|\beta}$.

Now let $\ell'\tau$ be a contrast of treatment effects and we are interested in testing

$$H_0: \ell'\tau = \ell_0, \quad H_A: \ell'\tau \neq \ell_0. \tag{2.25}$$

The bluc of $\ell'\tau$ is $\ell'\hat{\tau}$ where $\hat{\tau}$ is given by either expression of Theorem 2.3, and

$$\widehat{\mathrm{Var}}(\ell'\hat{\tau}) = \ell'(C_{\tau|\beta} + aJ_v)^{-1}\ell\hat{\sigma}^2, \quad \text{or} \quad \ell'\Omega\ell\hat{\sigma}^2. \tag{2.26}$$

Hence the test statistic for testing the null hypothesis of (2.25) is

$$t = \frac{\ell'\hat{\tau} - \ell_0}{\sqrt{\widehat{\mathrm{var}}(\ell'\hat{\tau})}}, \tag{2.27}$$

which is distributed as a t-variable with $\nu = n - b - v + 1$ degrees of freedom. The p-value for testing (2.25) is

$$p\text{-value} = 2P(t(\nu) > |t_{\mathrm{cal}}|), \tag{2.28}$$

where $t(\nu)$ is a t-variable with ν degrees of freedom and t_{cal} is the calculated t statistic of (2.27). P-value for a one-sided test of (2.25) can be easily calculated from the standard methods.

A $(1 - \alpha)100\%$ confidence interval for $\ell'\tau$ is

$$\ell'\hat{\tau} \pm \{t_{1-(\alpha/2)}(\nu)\}\sqrt{\hat{\mathrm{var}}(\ell'\hat{\tau})},$$

where $t_{1-(\alpha/2)}(\nu)$ is the $(1 - (\alpha/2))100$ percentile point of a t-distribution with ν degrees of freedom.

2.3 A Numerical Example

In this section we consider a numerical example with artificial data and explain the analysis discussed in Sec. 2.2.

Consider an experiment with $v = 4$, $b = 4$ and artificial data given in Table 2.3. We assume that the block effects are fixed effects. Here $\mathbf{r} = 4\mathbf{1}_4$, $\mathbf{k} = 4\mathbf{1}_4$,

$$N = \begin{bmatrix} 2 & 0 & 1 & 1 \\ 1 & 2 & 0 & 1 \\ 1 & 1 & 2 & 0 \\ 0 & 1 & 1 & 2 \end{bmatrix},$$

$\mathbf{T}' = (21, 34, 45, 52), \mathbf{B}' = (25, 39, 43, 45), G = 152, \mathrm{SS}_{\mathrm{T}} = 164, \mathrm{SS}_{\mathrm{Tr}} = 70.32,$ $\mathrm{SS}_{\mathrm{B}} = 61.0.$

$$C_{\tau|\beta} = 4\mathbf{I}_4 - \frac{1}{4}NN' = \frac{1}{4}\begin{bmatrix} 10 & -3 & -4 & -3 \\ -3 & 10 & -3 & -4 \\ -4 & -3 & 10 & -3 \\ -3 & -4 & -3 & 10 \end{bmatrix},$$

$$C_{\tau|\beta}^{-} = \frac{1}{42}\begin{bmatrix} 13 & 0 & 1 & 0 \\ 0 & 13 & 0 & 1 \\ 1 & 0 & 13 & 0 \\ 0 & 1 & 0 & 13 \end{bmatrix},$$

Table 2.3. Artificial Data in a block design experiment.

Block Number	Treatment (Response)			
1	A(3)	C(10)	A(4)	B(8)
2	B(8)	D(11)	C(12)	B(8)
3	C(12)	A(7)	C(11)	D(13)
4	D(15)	B(10)	A(7)	D(13)

$$\mathbf{Q}'_{\tau|\beta} = \left(\mathbf{T} - \frac{1}{k}N\mathbf{B}\right)' = (-13.5, -3, 7.5, 9).$$

Type III treatment SS $= \mathbf{Q}'_{\tau|\beta}C^{-}_{\tau|\beta}\mathbf{Q}_{\tau|\beta} = 28.37,$
Model SS $= 61 + 28.37 = 89.37,$
Error SS $= 164 - 89.37 = 74.63,$

$$C_{\beta|\tau} = 4\mathbf{I}_4 - \frac{1}{4}NN' = C_{\tau|\beta}.$$

Hence $C^{-}_{\beta|\tau} = C^{-}_{\tau|\beta}$

$$\mathbf{Q}'_{\beta|\tau} = \left(\mathbf{B} - \frac{1}{r}N\mathbf{T}\right)' = (-5.25, -2.25, 2.25, 5.25)$$

Type III block SS $= \mathbf{Q}'_{\beta|\tau}C^{-}_{\beta|\tau}\mathbf{Q}_{\beta|\tau} = 19.05.$

These results are summarized in the following tables:

ANOVA.

Source	df	SS	MS
Model	6	89.37	
Error	9	74.63	8.29

Type III sum of squares.

Source	df	SS	MS	F	p-value
Treatments	3	28.37	9.46	1.14	0.3842
Blocks	3	19.05	6.68	0.81	0.5196

In this example treatment effects as well as block effects are not significant at 0.05 level.

2.4 Analysis of Incomplete Block Designs with Recovery of Interblock Information

In this section we assume that the block sizes $k_1, k_2, \ldots, k_b$ are all equal and $k(< v)$ is the common block size. We further assume that the β_i of Model (2.1) are random effects following independent $N(0, \sigma_b^2)$ distribution and β_i are uncorrelated with random errors e_{ij}. Now

$$E(\hat{\sigma}^2) = EE(\hat{\sigma}^2|\beta_i \text{fixed}) = E(\sigma^2) = \sigma^2$$

$$E(\text{SS}_{\text{B}|\text{Tr}}) = EE[\{\mathbf{Q}'_{\beta|\tau} C^-_{\beta|\tau} Q_{\beta|\tau}\}|\beta_i \text{fixed}]$$

$$= EE[\text{tr}\{Q'_{\beta|\tau} C^-_{\beta|\tau} Q_{\beta|\tau}\}|\beta_i \text{fixed}]$$

$$= E[\text{tr}(\sigma^2 C_{\beta|\tau} C^-_{\beta|\tau} + C_{\beta|\tau}\boldsymbol{\beta}\boldsymbol{\beta}' C_{\beta|\tau} C^-_{\beta|\tau})]$$

$$= \sigma^2(b-1) + E[\text{tr}(\boldsymbol{\beta}\boldsymbol{\beta}' C_{\beta|\tau})]$$

$$= \sigma^2(b-1) + \sigma_b^2(\text{tr} C_{\beta|\tau})$$

$$= \sigma^2(b-1) + \sigma_b^2(bk-v).$$

Hence σ^2 is estimated by $\hat{\sigma}^2 = R_0^2/(n-b-v+1)$ and σ_b^2 is estimated by $\hat{\sigma}_b^2 = \{\text{SS}_{\text{B}|\text{Tr}} - (b-1)\hat{\sigma}^2\}/(bk-v)$.

In this case the treatment effects are estimated from within block and between block responses. Estimating τ from within block responses leads to the estimates given by Eq. (2.4). Further

$$E\left(\frac{1}{\sqrt{k}}\mathbf{B}\right) = \left[\sqrt{k}\mathbf{1}_b|\frac{1}{\sqrt{k}}\mathbf{N}'\right]\begin{bmatrix}\mu \\ \tau\end{bmatrix},$$

$$\text{Var}\left(\frac{1}{\sqrt{k}}\mathbf{B}\right) = (\sigma^2 + k\sigma_b^2)\,I_b. \tag{2.29}$$

Combining (2.4) and (2.29), we have the linear model

$$E\begin{bmatrix}\mathbf{Q}_{\tau|\beta} \\ \dfrac{1}{\sqrt{k}}\mathbf{B}\end{bmatrix} = \begin{bmatrix}0 & C_{\tau|\beta} \\ \sqrt{k}\mathbf{1}_b & \dfrac{1}{\sqrt{k}}N'\end{bmatrix}\begin{bmatrix}\mu \\ \tau\end{bmatrix},$$

$$\text{Var}\begin{bmatrix}\mathbf{Q}_{\tau|\beta} \\ \dfrac{1}{\sqrt{k}}\mathbf{B}\end{bmatrix} = \begin{bmatrix}\dfrac{1}{w}C_{\tau|\beta} & 0 \\ 0 & \dfrac{1}{w'}I_b\end{bmatrix}, \tag{2.30}$$

where $w = 1/\sigma^2$, $w' = 1/(\sigma^2 + k\sigma_b^2)$. The dispersion matrix in (2.30) is singular. The normal equations with singular dispersion matrix of the random observational vector discussed in Chap. 1 gives

$$\begin{bmatrix}bk & w'\mathbf{r}' \\ w'\mathbf{r} & wC_{\tau|\beta} + \dfrac{w'}{k}NN'\end{bmatrix}\begin{bmatrix}\tilde{\mu} \\ \tilde{\tau}\end{bmatrix} = \begin{bmatrix}w'G \\ w\mathbf{Q}_{\tau|\beta} + \dfrac{w'}{k}NB\end{bmatrix}, \tag{2.31}$$

where $\tilde{\mu}$ and $\tilde{\tau}$ are the estimators of μ and τ with recovery of interblock information. The coefficient matrix of the estimators in (2.31) is singular. Further, denoting the Kronecker product of matrices by $\otimes$, we have

$$wC_{\tau|\beta} + \frac{w'}{k}NN' = U'\{I_b \otimes (\sigma^2 I_k + \sigma_b^2 J_k)^{-1}\}U,$$

is non-singular, because U given by (2.2) is of rank v. Hence a solution of (2.31) is

$$\tilde{\mu} = 0, \quad \tilde{\tau} = \left(wC_{\tau|\beta} + \frac{w'}{k}NN'\right)^{-1}\left(wQ_{\tau|\beta} + \frac{w'}{k}N\mathbf{B}\right).$$

Since w and w' are unknown, we replace them by their estimates obtained from the estimates of σ^2 and σ_b^2 discussed earlier in this section to get

$$\tilde{\tau} = \left(\hat{w}C_{\tau|\beta} + \frac{\hat{w}'}{k}NN'\right)^{-1}\left(\hat{w}Q_{\tau|\beta} + \frac{\hat{w}'}{k}N\mathbf{B}\right),$$

$$\hat{\mathrm{Var}}(\tilde{\tau}) = \left(\hat{w}C_{\tau|\beta} + \frac{\hat{w}'}{k}NN'\right)^{-1}. \tag{2.32}$$

Since estimates of w and w' are used in $\tilde{\tau}$, the quantity $\tilde{\tau}'(\hat{w}Q_{\tau|\beta} + \frac{\hat{w}'}{k}N\mathbf{B})$ is approximately distributed as a $\chi^2(v-1)$ variable, and this statistic can be used to test the hypothesis (2.12).

Contrasts of treatment effects can be tested or estimated by confidence intervals from (2.32) assuming $\tilde{\tau}$ to have an approximate normal distribution.

Equations (2.31) can also be derived from the linear model (2.2), putting $\beta = 0$ identically and taking

$$\mathrm{Var}(\mathbf{Y}) = \{I_b \otimes (\sigma^2 I_k + \sigma_b^2 J_k)\}.$$

2.5 Nonparametric Analysis

We assume a general block design setting as indicated in Sec. 2.1. Let Y_{ij} be the response of the jth unit in the ith block and we assume that it has density in a location family

$$f_{ij}(y) = f(y - \beta_i - \tau_{d(i,j)}), \tag{2.33}$$

where $d(i, j)$ is the treatment applied to the jth unit in the ith block, β_i is the ith block effect and $\tau_{d(i,j)}$, is the $d(i, j)$ treatment effect. We assume no specific distributional assumptions while testing the null hypothesis

$$H_0(\mathrm{Tr}): \tau_1 = \tau_2 = \cdots = \tau_v, \qquad H_A(\mathrm{Tr}): \tau_i \neq \tau_j \quad \text{for some } i \neq j.$$

Under the null hypothesis, all the observations in the ith block are identically distributed. Hence we independently rank the observations in each of the b blocks. Let R_{ij} be the random variable denoting the rank assigned in the ith block for the response Y_{ij}. Then

$$P(R_{ij} = u) = \frac{1}{k_i}, \quad u = 1, 2, \ldots, k_i. \tag{2.34}$$

Hence

$$E(R_{ij}) = \sum_{u=1}^{k_i} \frac{u}{k_i} = (k_i + 1)/2, \tag{2.35}$$

$$\mathrm{Var}(R_{ij}) = \sum_{u=1}^{k_i} \frac{u^2}{k_i} - \left(\frac{k_i + 1}{2}\right)^2$$

$$= \frac{k_i^2 - 1}{12} = \frac{k_i - 1}{k_i}\sigma_i^2, \tag{2.36}$$

where $\sigma_i^2 = k_i(k_i + 1)/12$. Further for $j \neq j'$,

$$P(R_{ij} = u, R_{ij'} = w) = \frac{1}{k_i(k_i - 1)}, \quad u, w = 1, 2, \ldots, k_i; u \neq w. \tag{2.37}$$

Hence

$$\mathrm{Cov}(R_{ij}, R_{ij'}) = \sum_{\substack{u,w=1 \\ u \neq w}}^{k_i} \frac{uw}{k_i(k_i - 1)} - \left(\frac{k_i + 1}{2}\right)^2$$

$$= -\frac{(k_i + 1)}{12} = -\frac{\sigma_i^2}{k_i}. \tag{2.38}$$

Now, let

$$R_{i\ell}^* = \sum_{\substack{j=1 \\ d(i,j)=\ell}}^{k_i} R_{ij} - n_{i\ell}\left(\frac{k_i + 1}{2}\right). \tag{2.39}$$

Then $R_{i\ell}^*$ is the sum of the ranks measured from the block mid rank of the ℓth treatment in the ith block, and

$$E(R_{i\ell}^*) = 0, \quad \mathrm{Var}(R_{i\ell}^*) = \left\{n_{i\ell}\left(\frac{k_i - 1}{k_i}\right) - \left(\frac{n_{i\ell}(n_{i\ell} - 1)}{k_i}\right)\right\}\sigma_i^2$$

$$= \left(n_{i\ell} - \frac{n_{i\ell}^2}{k_i}\right)\sigma_i^2. \tag{2.40}$$

Also for $\ell \neq \ell'$

$$\text{Cov}(R_{i\ell}^*, R_{i\ell'}^*) = -\left(\frac{n_{i\ell} n_{i\ell'}}{k_i}\right) \sigma_i^2. \tag{2.41}$$

Put $R_\ell = \sum_{i=1}^{b} R_{i\ell}^*$, and $\mathbf{R}' = (R_1, R_2, \ldots, R_v)$. Then

$$E(\mathbf{R}) = \mathbf{0}, \quad \text{Var}(\mathbf{R}) = \Lambda = (\Lambda_{\ell\ell'}), \tag{2.42}$$

where

$$\Lambda_{\ell\ell} = \sum_{i=1}^{b} \left(n_{i\ell} - \frac{n_{i\ell}^2}{k_i}\right) \sigma_i^2, \quad \Lambda_{\ell\ell'} = -\sum_{i=1}^{b} \left(\frac{n_{i\ell} n_{i\ell'}}{k_i}\right) \sigma_i^2. \tag{2.43}$$

Assume that $\mathbf{R}$ has approximately a normal distribution and note that Λ is a singular matrix. Under the null hypothesis of equality of treatment effects, the statistic $T = \mathbf{R}' \Lambda^- \mathbf{R}$, where Λ^- is a symmetric g-inverse of Λ, has a $\chi^2 (v - 1)$ distribution. Hence the p-value for the test is

$$p\text{-value} = P(\chi^2(v - 1) > T_{\text{cal}}), \tag{2.44}$$

where T_{cal} is the calculated value of the test statistic $T = \mathbf{R}' \Lambda^- \mathbf{R}$.

When all block sizes are equal with common block size k, we have

$$\Lambda = \sigma^2 C_{\tau|\beta},$$

where

$$\sigma^2 = k(k + 1)/12.$$

For further details on nonparametric analysis of block designs, we refer to Desu and Raghavarao (2003).

2.6 Orthogonality

A design is said to be orthogonal for the estimation of treatment and block effects, if and only if

$$\text{Cov}(\ell' \hat{\tau}, \mathbf{m}' \hat{\boldsymbol{\beta}}) = 0, \tag{2.45}$$

where $\ell' \tau$ and $\mathbf{m}' \beta$ are estimable functions of treatment and block effects, respectively. The blues $\ell' \hat{\tau}$ and $\mathbf{m}' \hat{\boldsymbol{\beta}}$ are linear functions of $\mathbf{Q}_{\tau|\beta}$ and $\mathbf{Q}_{\beta|\tau}$, respectively. Hence (2.45) implies and is implied by

$$\text{Cov}(\mathbf{Q}_{\tau|\beta}, \mathbf{Q}_{\beta|\tau}) = 0. \tag{2.46}$$

Noting that

$$\mathrm{Cov}\left[\mathbf{T} - ND\left(\frac{1}{k_1}, \frac{1}{k_2}, \ldots, \frac{1}{k_b}\right)\mathbf{B}, \mathbf{B} - N'D\left(\frac{1}{r_1}, \frac{1}{r_2}, \ldots, \frac{1}{r_v}\right)\mathbf{T}\right]$$

$$= \sigma^2\left[N - N - N + ND\left(\frac{1}{k_1}, \frac{1}{k_2}, \ldots, \frac{1}{k_b}\right)N'D\left(\frac{1}{r_1}, \frac{1}{r_2}, \ldots, \frac{1}{r_v}\right)N\right]$$

$$= -\sigma^2 C_{\tau|\beta} D\left(\frac{1}{r_1}, \frac{1}{r_2}, \ldots, \frac{1}{r_v}\right)N,$$

(2.46) is equivalent to

$$C_{\tau|\beta} D\left(\frac{1}{r_1}, \frac{1}{r_2}, \ldots, \frac{1}{r_v}\right)N = 0. \tag{2.47}$$

For a connected design 0 is a simple root of $C_{\tau|\beta}$ with $\mathbf{1}_v$ as the associated eigenvector and (2.47) is equivalent to

$$D\left(\frac{1}{r_1}, \frac{1}{r_2}, \ldots, \frac{1}{r_v}\right)N = [a_1\mathbf{1}_v \quad a_2\mathbf{1}_v \quad \cdots \quad a_v\mathbf{1}_v] \tag{2.48}$$

for nonzero a_i's. Hence n_{ij}/r_i is constant for every j and

$$n_{ij} = r_i k_j/n, \quad \text{for every } i, j. \tag{2.49}$$

Condition (2.49) is called proportional cell frequencies. We thus proved

Theorem 2.4 *(B.V. Shah, 1959b) A connected block design is orthogonal for the estimation of treatment and block effects if and only if it has proportional cell frequencies, that is, $n_{ij} = r_i k_j/n$ for every $i = 1, 2, \ldots, v$; $j = 1, 2, \ldots, b$.*

The randomized block design with the $v \times b$ incidence matrix N consisting of all 1's is an orthogonal design for the estimation of treatment and block effects. For an orthogonal design, the sum of squares of treatments adjusted for blocks is the same as the treatment sum of squares ignoring blocks. The Type I and III sum of squares for blocks and treatments given in SAS package are same for an orthogonal design.

Let $\ell'\tau$ be any normalized contrast satisfying $\ell'\ell = 1$. The variance of its blue if a randomized block design of r blocks is used is $\sigma^2\ell'\ell/r = \sigma^2/r$ and Fisher information in the estimate is r/σ^2. With any other design in b blocks, and r replicates of each treatment, the information is $1/\{(\ell'C_{\tau|\beta}^-\ell)\sigma^2\}$. The relative loss of information in estimating $\ell'\tau$ by any block design instead of a randomized block design is

$$\frac{r - \dfrac{1}{\ell' C_{\tau|\beta}^{-} \ell}}{r}.$$

(2.50)

We now prove

Theorem 2.5 *Let ℓ_i be orthonormal eigenvectors corresponding to the nonzero eigenvalues of $C_{\tau|\beta}$ and let α_i be the relative loss of information in estimating $\ell'_i \tau$ for $i = 1, 2, \ldots, v - 1$. Then*

$$\sum_{i=1}^{v-1} \alpha_i = (b/r) - 1.$$

Proof. From (2.50)

$$\alpha_i = \left\{ r - \frac{\lambda_i^2}{\ell'_i C_{\tau|\beta} \ell_i} \right\} \Big/ r,$$

where λ_i is the eigenvalue of $C_{\tau|\beta}$ corresponding to the eigenvector ℓ_i. Now

$$\sum_{i=1}^{v-1} \alpha_i = v - 1 - \frac{1}{r} \sum_{i=1}^{v-1} \lambda_i = v - 1 - \frac{1}{r}(\operatorname{tr} C_{\tau|\beta})$$

$$= v - 1 - \frac{1}{r}(rv - b) = (b/r) - 1.$$

(2.51)

Theorem 2.5 implies that in a partially confounded factorial experiment where no main effect or interaction is totally confounded, the total relative loss of information in estimating main effects and interactions is one less than the number of blocks in a replication.

2.7 Variance and Combinatorial Balance

Several types of "balance" are used in statistical literature (see Preece, 1982). In block designs three types of balance are commonly used and they are

1. variance balance,
2. combinatorial balance, and
3. efficiency balance.

We will discuss the first two types in this section and discuss the third type in the next section. We have

Definition 2.3 A connected block design is said to be *variance balanced* if the variance of every estimated elementary contrast of treatment effects is the same.

Definition 2.4 A block design is said to be *combinatorially balanced* if $NN' = D + \lambda J_v$, where N is the incidence matrix of the design, D is a diagonal matrix, and λ is a positive integer.

For an equi-replicated, equi-block sized design, variance balance implies and is implied by combinatorial balance and such designs are known as balanced block designs. Balanced block designs with incomplete blocks are called balanced incomplete block designs and we will discuss them in detail in Chaps. 4 and 5.

When designs are not equi-replicated and equi-block sized, a block design may be variance balanced without being combinatorially balanced, and vice versa.

We first establish the characterization of $C_{\tau|\beta}$ for a variance balanced design. We prove

Theorem 2.6 *A connected block design is variance balanced if only if its $C_{\tau|\beta}$ is of the form*

$$C_{\tau|\beta} = \frac{2}{\bar{V}}\left(I_v - \frac{1}{v}J_v\right), \tag{2.52}$$

where $\bar{V}$ is the average variance of all estimated elementary contrasts of treatment effects.

Proof. Let $C_{\tau|\beta}^- = (C^{ij})$. The average variance of all estimated elementary contrasts of treatment effects, $\bar{V}$, is

$$\begin{aligned}
\bar{V} &= \frac{2}{v(v-1)}\sum_{i<j}\mathrm{var}(\hat{\tau}_i - \hat{\tau}_j) \\
&= \frac{2\sigma^2}{v(v-1)}\left[(v-1)\sum_i C^{ii} - \sum_{i\neq j}C^{ij}\right] \\
&= \frac{2\sigma^2}{v(v-1)}\left[v\,\mathrm{tr}(C_{\tau|\beta}^-) - \mathbf{1}_v' C_{\tau|\beta}^- \mathbf{1}_v\right]. \tag{2.53}
\end{aligned}$$

Assume that the connected block design is variance balanced so that

$$\mathrm{Var}(\hat{\tau}_i - \hat{\tau}_j) = \bar{V},$$

for every $i \neq j$. Also

$$\mathrm{Var}(\hat{\tau}_j - \hat{\tau}_\ell) = \mathrm{Var}[(\hat{\tau}_i - \hat{\tau}_\ell) - (\hat{\tau}_i - \hat{\tau}_j)],$$

so that,

$$\mathrm{Cov}(\hat{\tau}_i - \hat{\tau}_\ell, \hat{\tau}_i - \hat{\tau}_j) = \bar{V}/2.$$

Now

$$\mathrm{Var}\left[\frac{1}{\sqrt{\ell(\ell+1)}}\sum_{i=1}^{\ell}(\hat{\tau}_i - \hat{\tau}_{\ell+1})\right] = \frac{1}{\ell(\ell+1)}\left[\ell\bar{V} + \frac{\ell(\ell-1)}{2}\bar{V}\right] = \bar{V}/2,$$

$$\tag{2.54}$$

$$\mathrm{Cov}\left[\frac{1}{\sqrt{\ell(\ell+1)}}\sum_{i=1}^{\ell}(\hat{\tau}_i - \hat{\tau}_{\ell+1}), \frac{1}{\sqrt{\ell'(\ell'+1)}}\sum_{i=1}^{\ell'}(\hat{\tau}_i - \hat{\tau}_{\ell'+1})\right] = 0, \quad \ell \neq \ell'.$$

Without loss of generality, we prove the above covariance result for $\ell < \ell'$. Considering

$$\mathrm{Var}\left[\sum_{i=1}^{\ell}\left(\hat{\tau}_i - \hat{\tau}_{\ell+1}\right) - \sum_{i=1}^{\ell'}\left(\hat{\tau}_i - \hat{\tau}_{\ell'+1}\right)\right]$$

$$= \mathrm{Var}\left[(\ell+1)\left(\hat{\tau}_{\ell'+1} - \hat{\tau}_{\ell+1}\right) + \sum_{i=\ell+2}^{\ell'}\left(\hat{\tau}_{\ell'+1} - \hat{\tau}_i\right)\right]$$

and calculating the variances, we get

$$\frac{\ell(\ell+1)}{2}\bar{V} + \frac{\ell'(\ell'+1)}{2}\bar{V} - 2\mathrm{Cov}\left[\sum_{i=1}^{\ell}\left(\hat{\tau}_i - \hat{\tau}_{\ell+1}\right), \sum_{i=1}^{\ell'}\left(\hat{\tau}_i - \hat{\tau}_{\ell'+1}\right)\right]$$

$$= (\ell+1)^2\bar{V} + \frac{(\ell'-\ell-1)(\ell'-\ell)}{2}\bar{V} + (\ell+1)(\ell'-\ell-1)\bar{V},$$

from which we get the required covariance to be zero. Define a $(v-1) \times v$ matrix P by

$$P = \begin{pmatrix} \dfrac{1}{\sqrt{2}} & \dfrac{-1}{\sqrt{2}} & 0 & \cdots & 0 & 0 \\[2ex] \dfrac{1}{\sqrt{6}} & \dfrac{1}{\sqrt{6}} & \dfrac{-2}{\sqrt{6}} & \cdots & 0 & 0 \\[2ex] \vdots & \vdots & \vdots & \cdots & \vdots & \vdots \\[2ex] \dfrac{1}{\sqrt{v(v-1)}} & \dfrac{1}{\sqrt{v(v-1)}} & \dfrac{1}{\sqrt{v(v-1)}} & \cdots & \dfrac{1}{\sqrt{v(v-1)}} & \dfrac{-(v-1)}{\sqrt{v(v-1)}} \end{pmatrix}.$$

Then

$$PP' = I_{v-1}, \quad P'P = I_v - \frac{1}{v}J_v. \tag{2.55}$$

In view of (2.54), we have

$$PC^-_{\tau|\beta}P' = (\bar{V}/2)I_{v-1}$$

and consequently

$$P'(PC^-_{\tau|\beta}P')P = (\bar{V}/2)P'P,$$

$$\left(I_v - \frac{1}{v}J_v\right)C^-_{\tau|\beta}\left(I_v - \frac{1}{v}J_v\right) = (\bar{V}/2)\left(I_v - \frac{1}{v}J_v\right).$$

Pre- and post-multiplying the above equation by $C_{\tau|\beta}$, we get

$$C_{\tau|\beta}C^-_{\tau|\beta}C_{\tau|\beta} = (\bar{V}/2)C^2_{\tau|\beta}. \tag{2.56}$$

However, the left-hand side of (2.56) is $C_{\tau|\beta}$ and hence $C_{\tau|\beta} = (2/\bar{V})A$, where A is an idempotent matrix of rank $v-1$ and orthogonal to $\mathbf{1}_v$. This shows that $C_{\tau|\beta}$ is given by (2.52).

It can easily be verified that if $C_{\tau|\beta}$ satisfies (2.52), the design is variance balanced.

We can also characterize variance balance of connected block designs by the equality of nonzero eigenvalues of $C_{\tau|\beta}$ (see V. R. Rao, 1958; Raghavarao, 1971).

From the characterization given in Theorem 2.6, it can be verified that for an equi-replicated, equi-block sized connected variance balanced design.

$$NN' = aI_v + bJ_v \tag{2.57}$$

for suitable a and b, implying that the design is combinatorially balanced. In this class of designs,

$$\mathrm{tr}\left[\frac{2}{\bar{V}}\left(I_v - \frac{1}{v}J_v\right)\right] = \mathrm{tr}\left(rI_v - \frac{1}{k}NN'\right),$$

and if $n_{ij} = 1$ or 0, we have $\bar{V} = 2(v-1)/\{b(k-1)\}$. Equating $C_{\tau|\beta}$ to $\frac{2}{\bar{V}}\left(I_v - \frac{1}{v}J_v\right)$ and solving for NN', we get

$$NN' = \frac{r(v-k)}{v-1}I_v + \frac{r(k-1)}{v-1}J_v.$$

Since the entries of NN' are positive integers, $r(k-1)/(v-1)$ must be a positive integer, say λ, so that

$$NN' = (r-\lambda)I_v + \lambda J_v, \tag{2.58}$$

where $\lambda = r(k-1)/(v-1)$.

Designs whose NN' is a completely symmetric matrix are called Balanced Block Designs and were discussed by Shafiq and Federer (1979). These include the commonly used Randomized Block Designs, and Balanced Incomplete Block Designs. Another example of a Balanced Block Design with $v = 4 = b, r = 7 = k$ is

$$(1, 2, 3, 4, 1, 2, 3); \quad (1, 2, 3, 4, 1, 2, 4);$$
$$(1, 2, 3, 4, 1, 3, 4); \quad (1, 2, 3, 4, 2, 3, 4).$$

Designs with incidence matrix N and $k < v$ are called Balanced Incomplete Block (BIB) Designs. An example with $v = 4 = b, r = 3 = k, \lambda = 2$ is

$$(1, 2, 3); \quad (1, 2, 4); \quad (1, 3, 4); \quad (2, 3, 4).$$

Variance balanced designs were widely studied by Agarwal and Kumar (1984), Calvin (1986), Calvin and Sinha (1989), Gupta and Jones (1983), Gupta and Kageyama (1992), Gupta, Prasad and Das (2003), Jones, Sinha and Kageyama (1987), Kageyama (1989), Khatri (1982), Kulshresta, Dey and Saha (1972), Morgan and Uddin (1995), Sinha and Jones (1988) and Tyagi (1979).

Pairwise balanced designs introduced by Bose and Shrikhande (1960) in disproving Euler's conjecture on Orthogonal Latin Squares, and Symmetrical Unequal Block arrangements studied by Kishen (1940–1941), Raghavarao (1962a) are examples of combinatorially balanced designs.

The interconnection between these two types of balance was discussed by Hedayat and Federer (1974), and Hedayat and Stufken (1989).

The following design

$$(1, 2); (3, 4); (5, 6); (1, 3, 5); (1, 4, 6); (2, 3, 6); (2, 4, 5)$$

with $NN' = 2I_6 + J_6$ is combinatorially balanced, but is not variance balanced.

The following design

$$(0, 1); (0, 2); (0, 3); (0, 4); (1, 2, 3, 4); (1, 2, 3, 4), \qquad (2.59)$$

with $C_{\tau|\beta} = (5/2)(I_5 - (1/5)J_5)$ is variance balanced, but is not combinatorially balanced.

For balanced, equi-replicated designs it can be shown that N $D\left(\frac{1}{k_1}, \frac{1}{k_2}, \ldots, \frac{1}{k_b}\right) N'$ is non-singular and consequently (see Raghavarao, 1962b),

$$v \leq b. \qquad (2.60)$$

2.8 Efficiency Balance

We now turn our attention to Efficiency Balance and to this end, we define

Definition 2.5 (Pearce, 1968, 1970) The efficiency of a given design for an estimtated contrast of treatment effects $\ell'\tau$ is the ratio of its variance in an orthogonal design of the same experimental size to the variance in the given design.

We will use Ω as the g-inverse of $C_{\tau|\beta}$ in this discussion and the next section. Let

$$P = N\mathrm{diag}\left(\frac{1}{k_1}, \frac{1}{k_2}, \ldots, \frac{1}{k_b}\right)N' \tag{2.61}$$

and

$$M_0 = D\left(\frac{1}{r_1}, \frac{1}{r_2}, \ldots, \frac{1}{r_v}\right)P - \frac{1}{n}\mathbf{1r}', \tag{2.62}$$

so that

$$\Omega^{-1} = D(r_1, r_2, \ldots, r_v)[I_v - M_0]. \tag{2.63}$$

The efficiency in estimating $\ell'\tau$, denoted by $\mathrm{eff}(\ell'\tau)$, from Definition 2.5 is

$$\mathrm{Eff}(\ell'\tau) = \frac{\ell'D\left(\dfrac{1}{r_1}, \dfrac{1}{r_2}, \ldots, \dfrac{1}{r_v}\right)\ell}{\ell'[I_v - M_0]^{-1}D\left(\dfrac{1}{r_1}, \dfrac{1}{r_2}, \ldots, \dfrac{1}{r_v}\right)\ell}. \tag{2.64}$$

If $\mathrm{eff}(\ell'\tau)$, is taken as $1 - \mu$, it follows that

$$\ell'\left[D\left(\frac{1}{r_1}, \frac{1}{r_2}, \ldots, \frac{1}{r_v}\right) - (1 - \mu)(\mathbf{I}_v - \mathbf{M}_0)^{-1}\right.$$
$$\left. D\left(\frac{1}{r_1}, \frac{1}{r_2}, \ldots, \frac{1}{r_v}\right)\right]\ell = 0, \tag{2.65}$$

which holds if and only if

$$|\mathbf{M}_0 - \mu\mathbf{I}_v| = 0, \tag{2.66}$$

thereby implying that μ is an eigenvalue of M_0. We have thus established:

Theorem 2.7 *The efficiency of estimating a treatment contrast $\ell'\tau$ in a connected design is $1 - \mu$, where μ is an eigenvalue of M_0.*

Definition 2.6 A connected block design is *efficiency balanced*, if all the efficiencies in estimating treatment contrasts are the same.

Noting that

$$\mathbf{M}_0 \mathbf{1}_v = \mathbf{0} \tag{2.67}$$

as a consequence of Theorem 2.7, we have

Theorem 2.8 *A connected block design is efficiency balanced if and only M_0 is a zero matrix or M_0 has zero as an eigenvalue of multiplicity 1 and nonzero μ as an eigenvalue of multiplicity $v - 1$.*

When M_0 is a zero matrix, the block designs are trivially the randomized block designs with efficiency factor one for all estimated elementary contrasts. It can be verified that equi-replicated, equi-block sized, connected variance and combinatorial balanced designs are also efficiency balanced.

For the variance balanced design (2.59) given in the last section, we have

$$M_0 = \frac{1}{48} \begin{bmatrix} 12 & -31'_4 \\ -41_4 & 8I_4 - J_4 \end{bmatrix},$$

and its eigenvalues are 0, 1/3, 1/6, 1/6, 1/6. Hence the design (2.59) is not efficiency balanced.

Calinski (1971) gave the design

$$(1, 1, 2, 3); \ (1, 1, 2, 3); \ (1, 2); \ (1, 3); \ (2, 3) \tag{2.68}$$

for which

$$M_0 = \frac{1}{56} \begin{bmatrix} 4 & -2 & -2 \\ -3 & 5 & -2 \\ -3 & -2 & 5 \end{bmatrix},$$

with eigenvalues 0, 1/8, 1/8 and hence is efficiency balanced. It can be easily verified that $C_{\tau|\beta}$ for the design (2.68) is not completely symmetric and it is not variance balanced. Puri and Nigam (1975a) showed that a connected design is efficiency balanced if its P matrix given by (2.61) is of the form

$$P = D + p\mathbf{rr}', \tag{2.69}$$

where D is a diagonal matrix and p is a scalar.

For further results on efficiency balance we refer to Puri and Nigam (1975a,b). Interrelationships between the three types of efficiencies and some constructions are available in Puri and Nigam (1977b).

2.9 Calinski Patterns

Defining the matrices P and M_0 by Eqs. (2.61) and (2.62), the g-inverse of $C_{\tau|\beta}$, given by Ω can be expressed as

$$\Omega = D\left(\frac{1}{r_1}, \frac{1}{r_2}, \ldots, \frac{1}{r_v}\right) + \sum_{h=1}^{\infty} M_0^h D\left(\frac{1}{r_1}, \frac{1}{r_2}, \ldots, \frac{1}{r_v}\right). \qquad (2.70)$$

For a connected, efficiency balanced design (other than Randomized Block Designs) the nonzero eigenvalue of M_0 is μ, and $M_0 \mathbf{1}_v = \mathbf{0}, \mathbf{r}' M_0 = \mathbf{0}'$. Hence from spectral decomposition of asymmetric matrix, we can write

$$M_0 = \mu\left(I_v - \frac{1}{n}\mathbf{1}\mathbf{r}'\right). \qquad (2.71)$$

Clearly $M_0^h = \mu^{h-1} M_0$, from which it follows that

$$\Omega = (1-\mu)^{-1} D\left(\frac{1}{r_1}, \frac{1}{r_2}, \ldots, \frac{1}{r_v}\right) - \frac{\mu}{n(1-\mu)} J_v, \qquad (2.72)$$

and the solution of the normal equations estimating τ is

$$\hat{\tau} = \Omega \mathbf{Q}_{\tau|\beta} = (1-\mu)^{-1} D\left(\frac{1}{r_1}, \frac{1}{r_2}, \ldots, \frac{1}{r_v}\right) \mathbf{Q}_{\tau|\beta}. \qquad (2.73)$$

This results in a very simple form of statistical analysis for these designs. This simplicity is due to the fact that $M_0^h = \mu^{h-1} M_0$. Calinski noted that this relationship holds for a wider class of designs than just the efficiency balanced designs. The following theorem can be easily proved.

Theorem 2.9 *If for a given design there exists a set of $v-1$ linearly independent contrasts $\ell_i'\tau$ such that $v_1 (\leq v-1)$ of them satisfy*

$$M_0 \ell_i = \mu \ell_i, \quad i = 1, 2, \ldots, v_1$$

and the remaining $v - v_1$ of them satisfy

$$M_0 \ell_i = 0, \quad i = v_1 + 1, v_1 + 2, \ldots, v - 1,$$

then μ is the only nonzero eigenvalue of M_0 and

$$M_0 = \mu A, \qquad (2.74)$$

where A is an idempotent matrix of rank v_1. For such a design Ω will have the form given by Eq. (2.72).

We define

Definition 2.7 A design is said to have *Calinski pattern* (or *C*-pattern), if its M_0 matrix is a scalar multiple of an idempotent matrix, that is, has the form (2.74).

Calinski (1971) showed that the design

$$(1, 1, 2, 3); \quad (1, 1, 2, 3); \quad (2, 3) \tag{2.75}$$

has the M_0 matrix, $M_0 = \frac{1}{6}A$, where

$$A = \begin{pmatrix} 6/10 & -3/10 & -3/10 \\ -4/10 & 2/10 & 2/10 \\ -4/10 & 2/10 & 2/10 \end{pmatrix},$$

is an idempotent matrix of rank 1. Hence the design (2.75) has C-pattern. C-pattern designs can be augmented with more treatments in the blocks as given by the following theorem, whose proof is straightforward.

Theorem 2.10 *(Saha, 1976) Let N_i be the incidence matrix of a block design with parameters $v_i, b_i = b$, $\mathbf{r}_i$, $\mathbf{k}_i$ for $i = 1, 2$, where $\mathbf{r}_i$ and $\mathbf{k}_i$ are vectors of replication numbers and block sizes, respectively. Let $n_i = \mathbf{r}_i'\mathbf{1}_{v_i}$ and let N_1 has C-pattern with μ parameter μ_1. Then*

$$N = \begin{pmatrix} N_1 \\ N_2 \end{pmatrix}$$

is the incidence matrix of a C-pattern design if $N_2 = \mathbf{r}_2\mathbf{k}'/n$, where $\mathbf{k} = \mathbf{k}_1 + \mathbf{k}_2$ and $n = n_1 + n_2$. The μ parameter of the design with incidence matrix N is $\mu_1(n_1/n)$.

The specification on N_2 in Theorem 2.10 implies that $\mathbf{k}_2 = (n_2/n_1)\mathbf{k}_1$.

2.10 Optimality

Kiefer (1975b) consolidated and laid foundation to research on the concept of optimal designs. He developed the theory in terms of the information matrix $C_{\tau|\beta}$. For convenience we call $C_{\tau|\beta}$ simply by C-matrix and we use slightly different formulation given by Shah and Sinha (1989).

Let ϑ be a class of available designs with parameters $v, b, \mathbf{r}$ and $\mathbf{k}$. Let C_d be the C-matrix of the design $d \in \vartheta$. We suppress the subscript d as needed. We consider a class of optimality criteria Φ satisfying the following conditions:

(i) $\Phi(C_g)$ is the same for all g, where g is a permutation of $\{1, 2, \ldots, v\}$ and C_g is obtained from C by applying the permutation g to its rows and columns.

(ii) If C_1 and C_2 are two C-matrices of two designs such that $C_1 - C_2$ is nonnegative definite, then $\Phi(C_1) \leq \Phi(C_2)$.

(iii) For C_1 and C_2 as in (ii), $\Phi(C_1) \leq \Phi(C_2)$ if and only if $\Phi(tC_1) \leq \Phi(tC_2)$ for $t \geq 1$.

(iv) $\Phi\left(\sum C_g\right) \leq \Phi((v!)C)$ where $\sum C_g$ is overall $v!$ permutations g.

Condition (i) is motivated from the fact that the information matrix must be invariant for permutation of treatment labels. Condition (ii) makes the design with larger information optimal. If a design is optimal compared to another design, t copies of the first design must be optimal compared to t copies of the second design and this is reflected in (iii). Finally the information accrued by taking permutations of treatments should at least be as good as taking $v!$ copies of the original design and this is (iv).

We define

Definition 2.8   A design $d \in \vartheta$ is said to be *Extended Universal Optimal* in ϑ if its C-matrix minimizes every optimality functional Φ satisfying (i), (ii), (iii) and (iv).

We now prove

Theorem 2.11   *If $d^* \in \vartheta$ has a complete symmetric C_{d*} with maximum trace, then d^* is Extended Universal optimal in ϑ.*

Proof. Clearly $C_{d*} = a^*\left(I_v - \frac{1}{v}J_v\right)$, where $a^* = \mathrm{tr}(C_{d*})/(v - 1)$. Let d be any other design in ϑ. Then $\sum C_g$ can be represented as $a\left(I_v - \frac{1}{v}J_v\right)$, where $a = (v!)(\mathrm{tr}(C))/(v - 1)$. Since $\mathrm{tr}(C_{d*})$ is maximum, it follows that $a^*(v!) \geq a$. We also have

$$\sum C_g = a\left(I_v - \frac{1}{v}J_v\right) \leq a^*(v!)\left(I_v - \frac{1}{v}J_v\right) = v!C_{d*}$$

and by (ii)

$$\Phi\left(\sum C_g\right) \geq \Phi(v!C_{d*}). \tag{2.76}$$

Now suppose $\Phi(C_d) < \Phi(C_{d*})$. Then by (iv) and (ii)

$$\Phi\left(\sum C_g\right) \leq \Phi(v!C_d) < \Phi(v!C_{d*}),$$

contradicting (2.76), and proving the theorem.

Let $C_{\tau|\beta}$ be of rank $v - 1$ with nonzero eigenvalues $\lambda_1, \lambda_2, \ldots, \lambda_{v-1}$. Some of the interesting optimality functionals satisfying (i), (ii), (iii) and (iv) are:

- A-optimality: $\Phi(C) = \sum_{i=1}^{v-1} \frac{1}{\lambda_i}$. This minimizes the average variance of the estimated elementary contrasts.

- D-optimality: $\Phi(C) = \prod_{i=1}^{v-1} \frac{1}{\lambda_i}$. This minimizes the volume of the confidence ellipsoid of estimated orthonormal contrasts.
- E-optimality: maximize the smallest λ_i. This minimizes the maximum variance of any estimable normalized contrast.

For other generalizations and formulations, see Bagchi (1988), Bagchi and Bagchi (2001), Bondar (1983), Bueno Filho and Gilmour (2003), Cheng (1979, 1980, 1996), Cheng and Bailey (1991), Constantine (1981, 1982), Gaffke (1982), Jacroux (1980, 1983, 1984, 1985), Jacroux, Majumdar, and Shah (1995, 1997), John and Mitchell (1977), Magda (1979), Martin and Eccleston (1991), Shah and Sinha (2001), Takeuchi (1961), and Yeh (1986, 1988).

Kemthorne (1956) introduced the concept of efficiency factor, E, for a block design, given by

$$E = \frac{\bar{V}_0}{\bar{V}}, \tag{2.77}$$

where $\bar{V}_0(\bar{V})$ is the average variance of all elementary contrasts using an orthogonal design (current design) with the same experimental material. It can easily be shown that for equiblock sized connected designs

$$E = \frac{v(v-1)}{bk \sum (1/\lambda_i)},$$

where λ_i are eigenvalues of $C_{\tau|\beta}$.

Using the inequality on the arithmetic and harmonic means, we get

$$E \leq \frac{v \, \mathrm{tr}(C_{\tau|\beta})}{bk(v-1)}. \tag{2.78}$$

2.11 Transformations

The responses Y_{ij} are not always normally distributed so that the variances are not all equal and the tests are not valid. In such cases we can transform the Y_{ij} so that the transformed variables will have homoscedastic variances. These transformations, incidentally make the transformed variables nearly normally distributed in the transformed scale.

Let U be a random variable with $E(U) = \theta$, and $\mathrm{Var}(U) = a(\theta)$, a function of θ. We transform U to $W = h(U)$, so that $\mathrm{Var}(W)$ is a constant, σ^2, independent of θ. From the delta method, we have $\sigma^2 = \mathrm{Var}(W) = \{h'(\theta)\}^2 a(\theta)$, where $h'(\theta)$ is the derivative of $h(u)$ with respect to u evaluated at $u = \theta$ assuming $h'(\theta)$ exists and is not zero. Consequently

$$h(\theta) = \sigma \int \frac{1}{\sqrt{a(\theta)}} \, d\theta. \tag{2.79}$$

When Y_{ij} are binomially distributed with parameters m and p, $E(Y_{ij}/m) = p$, $\mathrm{Var}(Y_{ij}/m) = p(p-1)/m$, and from Eq. (2.79), we transform Y_{ij}/m to W_{ij} given by

$$W_{ij} = \sigma \int \frac{1}{\sqrt{p(p-1)}} \, dp \propto \sin^{-1} \sqrt{Y_{ij}/m}. \tag{2.80}$$

This transformation is known as Arc Sine transformation.

When Y_{ij} are distributed as a Poisson variable with parameter λ, $E(Y_{ij}) = \mathrm{Var}(Y_{ij}) = \lambda$, and the transformation is

$$W_{ij} = \sigma \int \frac{1}{\sqrt{\lambda}} \, d\lambda \propto \sqrt{Y_{ij}}. \tag{2.81}$$

This transformation is known as Square Root transformation.

When Y_{ij} are exponentially distributed with parameter θ so that $E(Y_{ij}) = 1/\theta$, $\mathrm{Var}(Y_{ij}) = 1/\theta^2$,

$$W_{ij} = \sigma \int \frac{1}{\sqrt{\theta^2}} \, d\theta \propto \ln(Y_{ij}). \tag{2.82}$$

This is known as Logarithmic transformation.

A wide class of transformations $W_{ij} = Y_{ij}^{\lambda}$ were considered by Box and Cox (1964) for appropriate λ value.

When Y_{ij} are binomial or multinomial probabilities, to restrict them to the (0,1) interval, we consider $Y_{ij} = e^{W_{ij}}/(1 + e^{W_{ij}})$, and consequently get the transformation

$$W_{ij} = \ln\left[Y_{ij}/(1 - Y_{ij})\right]. \tag{2.83}$$

This is called logit transformation and this is not a variance stabilizing transformation. Linear model will be considered for W_{ij}, and the analysis will be completed by the weighted least squares method. We will illustrate this method in Chap. 5.

The data are analyzed in the transformed scale and the conclusions drawn. Confidence intervals will be reported by reverting back to the original scale.

2.12 Covariance Analysis

We will now consider the covariance model

$$\mathbf{Y} = \begin{bmatrix} X & Z \end{bmatrix} \begin{bmatrix} \boldsymbol{\beta} \\ \boldsymbol{\gamma} \end{bmatrix} + \mathbf{e}, \tag{2.84}$$

where $\mathbf{Y}$ is an $n \times 1$ vector of response variables, X is an $n \times p$ design matrix with elements 0, 1, Z is an $n \times q$ matrix of quantitative variables, $\boldsymbol{\beta}$ is a vector of unknown parameters corresponding to the design factors, $\boldsymbol{\gamma}$ is a vector of regression coefficients and $\mathbf{e}$ is a vector of random errors assumed to have normal distribution with $E(\mathbf{e}) = \mathbf{0}$ and $\mathrm{Var}(\mathbf{e}) = \sigma^2 I_n$. We assume that rank of Z is q and the columns of Z are independent of the columns of X.

For the linear model $(\mathbf{Y}, X\boldsymbol{\beta}, \sigma^2 I_n)$, let $R_0^2 = \mathbf{Y}'A_0\mathbf{Y}$ be the minimum residual unconditional sum of square and $R_1^2 = \mathbf{Y}'A_1\mathbf{Y}$ be the minimum conditional residual sum of square to test the hypotheses $H_0\colon L'\boldsymbol{\beta} = \ell_0$, where L has s independent columns and all components of $L'\boldsymbol{\beta}$ are estimable.

The normal equations for estimating $\boldsymbol{\beta}$ and $\boldsymbol{\gamma}$ given by $\tilde{\boldsymbol{\beta}}$ and $\tilde{\boldsymbol{\gamma}}$ are

$$\begin{pmatrix} X'X & X'Z \\ Z'X & Z'Z \end{pmatrix} \begin{pmatrix} \tilde{\boldsymbol{\beta}} \\ \tilde{\boldsymbol{\gamma}} \end{pmatrix} = \begin{pmatrix} X'\mathbf{Y} \\ Z'\mathbf{Y} \end{pmatrix}$$

and eliminating $\tilde{\boldsymbol{\beta}}$, we get

$$Z'A_0Z\tilde{\boldsymbol{\gamma}} = Z'A_0\mathbf{Y}. \tag{2.85}$$

The unconditional minimum residual sum of squares for model (2.84) is

$$\tilde{R}_0^2 = R_0^2 - \tilde{\boldsymbol{\gamma}}'(Z'A_0\mathbf{Y})$$

with $n - r - q$ degrees of freedom, where r is the rank of X. Similarly, it can be shown that the conditional minimum residual sum of squares for testing $H_0\colon L'\boldsymbol{\beta} = \ell_o$, is

$$\tilde{R}_1^2 = R_1^2 - \tilde{\boldsymbol{\gamma}}^{*'}(Z'A_1\mathbf{Y})$$

with $n - r - q + s$ degrees of freedom, where $s = \mathrm{rank}(L)$, and $\tilde{\boldsymbol{\gamma}}^*$ satisfy a similar equation as (2.85) replacing A_0 by A_1. The test statistic for testing $H_0\colon L'\boldsymbol{\beta} = \ell_0$, is

$$F = \frac{\left(\tilde{R}_1^2 - \tilde{R}_0^2\right)/s}{\tilde{R}_0^2/(n - r - q)}, \tag{2.86}$$

which is distributed as an F-ratio with s numerator and $n - r - q$ denominator degrees of freedom.

This covariance analysis can be applied to a general block design setting as indicated at the beginning of this chapter. We consider one covariate and assume that its value is not affected by the block or treatment effects. We assume

$$Y_{ij} = \mu + \beta_i + \tau_{d(i,j)} + z_{ij}\gamma + e_{ij}. \tag{2.87}$$

Let $\mathbf{T}^z$, $\mathbf{B}^z$ be the vectors of treatment and block totals for the covariate Z and $\mathbf{Q}^z_{\tau|\beta} = \mathbf{T}^z - N\mathrm{diag}\left(\frac{1}{k_1}, \frac{1}{k_2}, \ldots, \frac{1}{k_b}\right)\mathbf{B}^z$. Then $\tilde{\gamma}$ can be estimated from

$$\left\{ \sum_{i,j} z_{ij}^2 - \sum_i \frac{z_{i\bullet}^2}{k_i} - \mathbf{Q}^{z'}_{\tau|\beta} C^-_{\tau|\beta} \mathbf{Q}^z_{\tau|\beta} \right\} \tilde{\gamma}$$

$$= \left[\sum_{i,j} z_{ij} Y_{ij} - \sum_i \frac{z_{i\bullet} Y_{i\bullet}}{k_i} - \mathbf{Q}'_{\tau|\beta} C^-_{\tau|\beta} \mathbf{Q}^z_{\tau|\beta} \right] \tag{2.88}$$

and the unconditional residual sum of squares is

$$\tilde{R}_0^2 = \sum_{i,j} Y_{ij}^2 - \sum_i Y_{i\bullet}^2 - \mathbf{Q}'_{\tau|\beta} C^-_{\tau|\beta} \mathbf{Q}_{\tau|\beta}$$

$$- \tilde{\gamma} \left[\sum_{i,j} z_{ij} Y_{ij} - \sum_i \frac{z_{i\bullet} Y_{i\bullet}}{k_i} - \mathbf{Q}'_{\tau|\beta} C^-_{\tau|\beta} \mathbf{Q}^z_{\tau|\beta} \right]. \tag{2.89}$$

In Eq. (2.88), note that $Y_{i\bullet} = \sum_j Y_{ij}$, and $z_{i\bullet} = \sum_j z_{ij}$.

Conditional on $H_0(\mathrm{Tr})$ of equal treatment effects, the conditional minimum residual sum of squares is

$$\tilde{R}_1^2 = \left[\sum_{i,j} Y_{ij}^2 - \sum_i \frac{Y_{i\bullet}^2}{k_i} \right] - \left[\left\{ \sum_{i,j} z_{ij} Y_{ij} \right. \right.$$

$$\left. \left. - \sum_i \frac{z_{i\bullet} Y_{i\bullet}}{k_i} \right\}^2 \Big/ \left\{ \sum_{i,j} z_{ij}^2 - \sum_i \frac{z_{i\bullet}^2}{k_i} \right\} \right].$$

The test statistic for testing $H_0(\mathrm{Tr})$ is then (2.86) with numerator degrees of freedom $v-1$ and denominator degrees of freedom $n-v-b$. It may be noted that $(X'X)\tilde{\beta} = X'Y - X'Z\tilde{\gamma}$ for model (2.84) and estimated dispersion matrix of $\tilde{\gamma}$ is

$$\hat{\mathrm{Var}}(\tilde{\gamma}) = \frac{\tilde{R}_0^2}{n-r-q}(Z'A_0Z)^{-1}, \tag{2.90}$$

and this can be used to estimate $\text{Var}(\tilde{\beta})$, which can be used to test and set confidence interval for contrasts of treatment effects.

Covariance analysis is used in an interesting way to adjust for interplot competition in David, Monod and Amoussou (2000).

3

Randomized Block Designs

3.1 Analysis with Fixed Block Effects

Randomized Block Design (RBD) is the simplest and commonly used design for experimental purposes. The experimental material consists of b blocks each of v units and v treatments are applied so that each treatment occurs exactly once in each block. In the notation of Chap. 2,

$$\mathbf{r} = b\mathbf{1}_v, \quad \mathbf{k} = v\mathbf{1}_b, \quad N = J_{vb}, \tag{3.1}$$

where J_{vb} is a $v \times b$ matrix of all 1's. Also,

$$C_{\tau|\beta} = b\left(I_v - \frac{1}{v}J_v\right), \quad C_{\beta|\tau} = v\left(I_b - \frac{1}{b}J_b\right), \tag{3.2}$$

where I_v and J_v are $v \times v$ identity matrix and matrix of all 1's, respectively. The g-inverses of the C-matrices are

$$C_{\tau|\beta}^{-} = \frac{1}{b}I_v, \quad C_{\beta|\tau}^{-} = \frac{1}{v}I_b. \tag{3.3}$$

Without loss of generality, let Y_{ij} be the response from the ith block receiving the jth treatment for $i = 1, 2, \ldots, b; \ j = 1, 2, \ldots, v$. Then the treatment total $T_j = \sum_i Y_{ij} = Y_{\bullet j}$ and the block total $B_i = \sum_j Y_{ij} = Y_{i\bullet}$. The adjusted treatment and block totals are respectively

$$\mathbf{Q}'_{\tau|\beta} = \left[Y_{\bullet 1} - \frac{1}{v}Y_{\bullet\bullet}, \ Y_{\bullet 2} - \frac{1}{v}Y_{\bullet\bullet}, \ldots, Y_{\bullet v} - \frac{1}{v}Y_{\bullet\bullet}\right],$$

$$\mathbf{Q}'_{\beta|\tau} = \left[Y_{1\bullet} - \frac{1}{b}Y_{\bullet\bullet}, \ Y_{2\bullet} - \frac{1}{b}Y_{\bullet\bullet}, \ldots, Y_{b\bullet} - \frac{1}{b}Y_{\bullet\bullet}\right], \tag{3.4}$$

where $Y_{\bullet\bullet} = \sum_{i,j} Y_{ij}$, is the grand total of all responses. The solutions of the normal equations are

$$\hat{\tau}_j = \bar{Y}_{\bullet j} - \bar{Y}_{\bullet\bullet}, \quad \hat{\beta}_i = \bar{Y}_{i\bullet} - \bar{Y}_{\bullet\bullet}, \tag{3.5}$$

where $\bar{Y}_{\bullet j} = Y_{\bullet j}/b, \ \bar{Y}_{i\bullet} = Y_{i\bullet}/v, \ \bar{Y}_{\bullet\bullet} = Y_{\bullet\bullet}/(vb)$.

Here

$$SS_{Tr} = SS_{Tr|B} = \sum_j \frac{Y_{\bullet j}^2}{b} - \frac{Y_{\bullet\bullet}^2}{vb},$$

$$SS_B = SS_{B|Tr} = \sum_i \frac{Y_{i\bullet}^2}{v} - \frac{Y_{\bullet\bullet}^2}{vb}. \tag{3.6}$$

Finally

$$R_0^2 = \left(\sum_{i,j} Y_{ij}^2 - \frac{Y_{\bullet\bullet}^2}{vb} \right) - \left(\sum_j \frac{Y_{\bullet j}^2}{b} - \frac{Y_{\bullet\bullet}^2}{vb} \right) - \left(\sum_i \frac{Y_{i\bullet}^2}{v} - \frac{Y_{\bullet\bullet}^2}{vb} \right)$$

$$= \sum_{i,j} \left(Y_{ij} - \bar{Y}_{\bullet j} - \bar{Y}_{i\bullet} + \bar{Y}_{\bullet\bullet} \right)^2 . \tag{3.7}$$

The ANOVA and Type III Sum of Squares are given in Tables 3.1 and 3.2.

The null hypothesis $H_0(\text{Tr})$: $\tau_1 = \tau_2 = \cdots = \tau_v$ is rejected in favor of the alternative hypothesis $H_A(\text{Tr})$: $\tau_j \neq \tau_{j'}$ for at least one pair (j, j'), $j \neq j'$ if p_1 of Table 3.2 is less than the significance level α. Similarly, the null hypothesis $H_0(B)$: $\beta_1 = \beta_2 = \cdots = \beta_b$ is rejected in favor of the alternative hypothesis $H_A(B)$: $\beta_i \neq \beta_{i'}$ for at least one pair of (i, i'), $i \neq i'$, if p_2 of Table 3.2 is less than the significance level α.

Table 3.1. ANOVA.

Source	df	SS	MS
Model	$v + b - 2$	$\sum_j \frac{Y_{\bullet j}^2}{b} + \sum_i \frac{Y_{i\bullet}^2}{v} - 2\frac{Y_{\bullet\bullet}^2}{vb}$	
Error	$(v-1)(b-1)$	$\sum_{i,j} Y_{ij}^2 - \sum_j \frac{Y_{\bullet j}^2}{b} - \sum_i \frac{Y_{i\bullet}^2}{v} + \frac{Y_{\bullet\bullet}^2}{vb}$	$\hat{\sigma}^2$

Table 3.2. TYPE III sum of squares.

Source	df	SS	MS	F	p
Treatments	$v-1$	$\sum_j \frac{Y_{\bullet j}^2}{b} - \frac{Y_{\bullet\bullet}^2}{vb}$	MS_{Tr}	$\frac{MS_{Tr}}{\hat{\sigma}^2} = F_{Tr}$	p_1^*
Blocks	$b-1$	$\sum_i \frac{Y_{i\bullet}^2}{v} - \frac{Y_{\bullet\bullet}^2}{vb}$	MS_B	$\frac{MS_B}{\hat{\sigma}^2} = F_B$	p_2^+

$p_1^* = P(F(v-1, (v-1)(b-1)) \geq F_{Tr}(\text{cal}))$, $p_2^+ = P(F(b-1, (v-1)(b-1)) \geq F_B(\text{cal}))$. $F_{Tr}(\text{cal})$, $F_B(\text{cal})$ are calculated F statistics F_{Tr} and F_B respectively, $F(a, b)$ is the F variable with numerator df of a and denominator df of b.

The treatment contrast, $\sum_j \ell_j \tau_j$ such that $\sum_j \ell_j = 0$, is different from ℓ_0 can be concluded by testing the null hypothesis

$$H_0: \sum_j \ell_j \tau_j = \ell_0, \quad H_A: \sum_j \ell_j \tau_j \neq \ell_0, \tag{3.8}$$

using the test statistic

$$t = \frac{\sum_j \ell_j \bar{Y}_{\bullet j} - \ell_0}{\sqrt{\hat{\sigma}^2 \left(\sum_j \ell_j^2 / b \right)}} \tag{3.9}$$

and

$$p\text{-value} = 2P(t((v-1)(b-1)) > |t(\text{cal})|) < \alpha. \tag{3.10}$$

In Eq. (3.10), $t((v-1)(b-1))$ is the t-variable with $(v-1)(b-1)$ degrees of freedom, $t(\text{cal})$ is the calculated t-statistic of (3.9) and α is the selected level of significance. A $(1-\alpha)100\%$ confidence interval for the treatment contrast $\sum_j \ell_j \tau_j$ is

$$\sum_j \ell_j \bar{Y}_{\bullet j} \pm t_{1-(\alpha/2)}((v-1)(b-1)) \sqrt{\hat{\sigma}^2 \left(\sum_j \ell_j^2 / b \right)}, \tag{3.11}$$

where $t_{1-(\alpha/2)}((v-1)(b-1))$ is the $(1-(\alpha/2))100$ percentile point of the t-distribution with $(v-1)(b-1)$ degrees of freedom.

Let $\mathbf{Y}' = (Y_{11}, Y_{12}, \ldots, Y_{1v}, Y_{21}, Y_{22}, \ldots, Y_{2v}, \ldots, Y_{b1}, Y_{b2}, \ldots, Y_{bv})$ be the vector of all responses arranged by treatments in the blocks. Then in matrix form, different sums of squares are given by quadratic forms in Y as follows:

$$
\begin{aligned}
\text{SS}_\text{T} &= \mathbf{Y}' \left[I_{vb} - \frac{1}{vb} J_{vb} \right] \mathbf{Y}, \\
\text{SS}_\text{Tr} &= \mathbf{Y}' \left[\frac{1}{b} J_b \otimes \left(I_v - \frac{1}{v} J_v \right) \right] \mathbf{Y}, \\
\text{SS}_\text{B} &= \mathbf{Y}' \left[\left(I_b - \frac{1}{b} J_b \right) \otimes \frac{1}{v} J_v \right] \mathbf{Y}, \\
R_0^2 &= \mathbf{Y}' \left[\left(I_b - \frac{1}{b} J_b \right) \otimes \left(I_v - \frac{1}{v} J_v \right) \right] \mathbf{Y}.
\end{aligned}
\tag{3.12}
$$

In (3.12), $\otimes$ denotes the Knocker product of matrices. Note that the matrices of quadratic forms of SS_Tr, SS_B, R_0^2, are idempotents. We need the representations of (3.12) in the next section to derive the distribution of the test statistic when block effects are random.

Sometimes the experimenter will be interested in judging the significance of all possible elementary contrasts of treatment effects of the type $\tau_j - \tau_{j'}$ for j, $j' = 1, 2, \ldots, v$; $j \neq j'$. If $v(v-1)/2$ tests are performed using the t-statistics of (3.9), it is possible to find at least one significant pair in an experiment, where all treatments have equal effects, when v is large. This is because, the t test controls the error rate contrast-wise and when $v(v-1)/2$ comparisons are made, with large v, at least one comparison is significant when H_0 is true.

In this case, experimenters use experimental-wise (family-wise) error rate of α, which means that on an average in $100\alpha\%$ experiments, when v treatments have equal effects, at least one pair of treatments will be concluded to have significant effect. To this end, we note that

$$\mathrm{Var}\left(\frac{\sqrt{b}}{\sigma}\left(\bar{Y}_{\bullet j} - \bar{Y}_{\bullet j'}\right)\right) = 2 = \mathrm{Var}(Z_j - Z_{j'}), \quad j, j' = 1, 2, \ldots, v; \quad j \neq j', \tag{3.13}$$

where Z_j and $Z_{j'}$ are independent standard normal variables. Hence treatments j and j' are significantly different if,

$$\left|\bar{Y}_{\bullet j} - \bar{Y}_{\bullet j'}\right| > \{q_{1-\alpha}(v, \infty)\}\frac{\sigma}{\sqrt{b}}, \tag{3.14}$$

where $q_{1-\alpha}(v, \infty)$ is the $(1-\alpha)100$ percentile of the Studentized Range distribution of v independent $N(0, 1)$ variables. Since σ is unknown, we replace σ^2 by $\hat{\sigma}^2$ and replace $q_{1-\alpha}(v, \infty)$ by $q_{1-\alpha}(v, (v-1)(b-1))$, where $q_{1-\alpha}(v, (v-1)(b-1))$ is the percentile point of the Studentized Range distribution estimating σ^2 by $\hat{\sigma}^2$ with $(v-1)(b-1)$ degrees of freedom.

Thus the treatments j and j' are concluded to be significant if

$$\left|\bar{Y}_{\bullet j} - \bar{Y}_{\bullet j'}\right| > \{q_{1-\alpha}(v, (v-1)(b-1))\}\sqrt{\frac{\hat{\sigma}^2}{b}}, \tag{3.15}$$

for $j, j' = 1, 2, \ldots, v$; $j \neq j'$. This is known as Tukey's Range Comparison Test. Another commonly used multiple comparison method for testing not only elementary contrasts, but also any of several contrasts is Scheffe's test. Here any contrast $\sum_j \ell_j \tau_j$ is significant, if

$$\left|\sum_j \ell_j \bar{Y}_{\bullet j}\right| > \sqrt{(v-1)F_{1-\alpha}(v-1, (v-1)(b-1))}\sqrt{\hat{\sigma}^2\left(\sum_j \left(\ell_j^2/b\right)\right)}, \tag{3.16}$$

where $F_{1-\alpha}(v-1, (v-1)(b-1))$ is the $(1-\alpha)100$ percentile point of $F(v-1, (v-1)(b-1))$ distribution.

There is a vast literature on multiple comparisons, and the interested reader is referred to Hochberg and Tamhane (1987) and Benjamini and Hochberg (1995).

3.2 Analysis with Random Block Effects

In this section we assume that the blocks are randomly selected from a population of available blocks. Then in the model

$$Y_{ij} = \mu + \beta_i + \tau_j + e_{ij}, \tag{3.17}$$

we assume β_i to be independent $N(0, \sigma_b^2)$ variables, whereas e_{ij} are independent $N(0, \sigma^2)$ variables. We further assume that β_i and e_{ij} are independently distributed. We now have

$$\begin{aligned}
\text{Var}(\mathbf{Y}) &= I_b \otimes \left(\sigma^2 I_v + \sigma_b^2 J_v\right) \\
&= V \,(\text{say}).
\end{aligned} \tag{3.18}$$

We verify that

$$\frac{1}{\sigma^2}\left[\left(I_b - \frac{1}{b}J_b\right) \otimes \left(I_v - \frac{1}{v}J_v\right)\right]\left[\left(I_b \otimes \left(\sigma^2 I_v + \sigma_b^2 J_v\right)\right)\right]$$

$$\frac{1}{\sigma^2}\left[\left(I_b - \frac{1}{b}J_b\right) \otimes \left(I_v - \frac{1}{v}J_v\right)\right]\left[\left(I_b \otimes \left(\sigma^2 I_v + \sigma_b^2 J_v\right)\right)\right]$$

$$= \left[\left(I_b - \frac{1}{b}J_b\right) \otimes \left(I_v - \frac{1}{v}J_v\right)\right]\left[\left(I_b - \frac{1}{b}J_b\right) \otimes \left(I_v - \frac{1}{v}J_v\right)\right]$$

$$= \left[\left(I_b - \frac{1}{b}J_b\right) \otimes \left(I_v - \frac{1}{v}J_v\right)\right]$$

$$= \frac{1}{\sigma^2}\left[\left(I_b - \frac{1}{b}J_b\right) \otimes \left(I_v - \frac{1}{v}J_v\right)\right]\left[\left(I_b \otimes \left(\sigma^2 I_v + \sigma_b^2 J_v\right)\right)\right]. \tag{3.19}$$

Hence from the distribution theory given in Sec. 1.6, $R_0^2/\sigma^2 \sim \chi^2\,((v-1)(b-1))$. Also,

$$\frac{1}{\sigma^2}\left[\frac{1}{b}J_b \otimes \left(I_v - \frac{1}{v}J_v\right)\right]\left[\left(I_b \otimes \left(\sigma^2 I_v + \sigma_b^2 J_v\right)\right)\right]$$

$$\frac{1}{\sigma^2}\left[\frac{1}{b}J_b \otimes \left(I_v - \frac{1}{v}J_v\right)\right]\left[\left(I_b \otimes \left(\sigma^2 I_v + \sigma_b^2 J_v\right)\right)\right]$$

$$= \left[\frac{1}{b}J_b \otimes \left(I_v - \frac{1}{v}J_v\right)\right]\left[\frac{1}{b}J_b \otimes \left(I_v - \frac{1}{v}J_v\right)\right]$$

$$= \left[\frac{1}{b}J_b \otimes \left(I_v - \frac{1}{v}J_v\right)\right]$$

$$= \frac{1}{\sigma^2}\left[\frac{1}{b}J_b \otimes \left(I_v - \frac{1}{v}J_v\right)\right]\left[\left(I_b \otimes \left(\sigma^2 I_v + \sigma_b^2 J_v\right)\right)\right]. \tag{3.20}$$

Hence $SS_{Tr}/\sigma^2 \sim \chi^2(v-1)$. Also,

$$\left[\left(I_b - \frac{1}{b}J_b\right) \otimes \left(I_v - \frac{1}{v}J_v\right)\right]\left[(I_b \otimes (\sigma^2 I_v + \sigma_b^2 J_v))\right]\left[\frac{1}{b}J_b \otimes \left(I_v - \frac{1}{v}J_v\right)\right] = 0$$

and SS_{Tr} and R_0^2 in this case are independently distributed.

Thus the F statistic given in Table 3.2 for testing the equality of treatment effects is still valid when the block effects are random and the dispersion matrix of the response vector is of the form (3.18). Huynh and Feldt (1970), noted that the F-statistic given in Table 3.2 for testing the treatment effects is still valid if

$$\text{Var}(\mathbf{Y}) = I_b \otimes \Sigma,$$

where Σ is of the form

$$\Sigma = dI_v + \mathbf{u}\mathbf{1}_v' + \mathbf{1}_v\mathbf{u}', \tag{3.21}$$

where $\mathbf{u}$ is any arbitrary vector, and d is a positive scalar. Further more, $R_0^2/d \sim \chi^2$ $((v-1)(b-1))$ and $SS_{Tr}/d \sim \chi^2(v-1)$, and both are independently distributed.

3.3 Unequal Error Variances

In some occasions, we may have fixed block effects; but error variances may be different from block to block. In the model

$$Y_{ij} = \mu + \beta_i + \tau_j + e_{ij},$$

we assume that $\text{Var}(e_{ij}) = \sigma_i^2$, for every i and j. Let $\hat{e}_{ij} = Y_{ij} - \bar{Y}_{i\bullet} - \bar{Y}_{\bullet j} + \bar{Y}_{\bullet\bullet}$ be the residual from the response Y_{ij}. Let $S_i = \sum_j \hat{e}_{ij}^2$.

Then it can be easily verified that

$$E(S_i) = \frac{(v-1)(b-2)}{b}\sigma_i^2 + \frac{(v-1)}{b^2}\sum_i \sigma_i^2, \quad i = 1, 2, \ldots, b. \tag{3.22}$$

Let $\mathbf{S}' = (S_1, S_2, \ldots, S_b)$ and $\boldsymbol{\sigma}' = \left(\sigma_1^2, \sigma_2^2, \ldots, \sigma_b^2\right)$. Equation (3.22) can be written in matrix form as

$$E(\mathbf{S}) = \left(\frac{(v-1)(b-2)}{b}I_b + \frac{(v-1)}{b^2}J_b\right)\boldsymbol{\sigma}. \tag{3.23}$$

Hence $\boldsymbol{\sigma}$ is unbiasedly estimated by $\hat{\boldsymbol{\sigma}}$:

$$\hat{\boldsymbol{\sigma}} = \frac{b}{(v-1)(b-2)}\left[I_b - \frac{1}{b(b-1)}J_b\right]\mathbf{S}. \tag{3.24}$$

Note that $R_0^2 = \sum_{i=1}^{b} S_i$, and

$$E(R_0^2) = \frac{(v-1)(b-1)}{b} \sum_i \sigma_i^2. \tag{3.25}$$

In this case we test the equality of treatment effects by using the F-statistic of Table 3.2 with 1 df for numerator and b df for denominator.

There is an extensive literature on this topic and the interested reader is referred to Brindley and Bradley (1985), Ellenberg (1977), Grubbs (1948), Maloney (1973) and Russell and Bradley (1958).

3.4 Permutation Test

In some cases the response variable may not be normally distributed. Then the equality of treatment effects can be tested by nonparametric methods ranking the responses in each block separately and the associated test is called Friedman's test (1937).

Alternatively, one may use permutation test which is also known as randomization test. Since the treatments are randomly assigned to each block, if the treatment effects are all the same, then the observed responses might have come from any treatment, not necessarily the treatment applied to the unit. With this in mind, we make all possible $(v!)^b$-arrangements and calculate a reasonable test statistic from each configuration. Usually the test statistic is the parametric analogue, or, equivalently

$$T = \sum_j \frac{T_j^2}{b}, \tag{3.26}$$

where T_j is the jth treatment total. The distribution of T will be generated based on $(v!)^b$ configurations of data and the p-value calculated from the generated distribution as

$$p\text{-value} = (\# \text{ configurations with } T > T(\text{cal}))/(v!)^b,$$

where $T(\text{cal})$ is the calculated T of (3.26) for the observed configuration of data. For further details see Edington (1995) or Good (2000).

3.5 Treatment Block Interactions

If the experimenter wishes to test the interaction of treatments and blocks, either the design can be modified by using each treatment in each block d (>1) times, or the analysis can be modified by creating a single degree of freedom

Table 3.3. Anova of generalized randomized block design.

Source	df	SS	MS
Model	$vb - 1$	$\sum_{i,j} \dfrac{Y_{ij\bullet}^2}{d} - \dfrac{Y_{\bullet\bullet\bullet}^2}{vbd}$	
Error	$vb(d - 1)$	$\sum_{i,j,l} Y_{ijl}^2 - \sum_{i,j} \dfrac{Y_{ij\bullet}^2}{d}$	$\hat{\sigma}^2$

Table 3.4. Type III sum of squares for generalized randomized block design.

Source	df	SS	MS	F	p
Treatments	$v - 1$	$\sum_j \dfrac{Y_{\bullet j\bullet}^2}{bd} - \dfrac{Y_{\bullet\bullet\bullet}^2}{vbd}$	MS_{Tr}	$\dfrac{MS_{Tr}}{\hat{\sigma}^2} = F_{Tr}$	p_1^*
Blocks	$b - 1$	$\sum_i \dfrac{Y_{i\bullet\bullet}^2}{vd} - \dfrac{Y_{\bullet\bullet}^2}{vbd}$	MS_B	$\dfrac{MS_B}{\hat{\sigma}^2} = F_B$	p_2^+
Tr $\times$ Bl	$(v - 1)(b - 1)$	$\sum_{i,j} \dfrac{Y_{ij\bullet}^2}{d} - \sum_j \dfrac{Y_{\bullet j\bullet}^2}{bd}$ $- \dfrac{Y_{i\bullet\bullet}^2}{vd} + \dfrac{Y_{\bullet\bullet\bullet}^2}{vbd}$	MS_I	$\dfrac{MS_I}{\hat{\sigma}^2} = F_I$	p_3^{++}

$p_1^* = P(F(v - 1, vb(d - 1)) \geq F_{Tr}(\text{cal})),\ p_2^+ = P(F(b - 1, vb(d - 1) \geq F_B(\text{cal})),\ p_3^{++} = P(F((v - 1)(b - 1), vb(d - 1)) > F_I(\text{cal}))$

sum of squares for non-additivity. The modified design may be called generalized randomized block design. Let Y_{ijl} be the response from ℓth unit in the ith block receiving jth treatment, for $\ell = 1, 2, \ldots, d; i = 1, 2, \ldots, b; j = 1, 2, \ldots, v$. Let $Y_{ij\bullet} = \sum_l Y_{ijl}$, $Y_{i\bullet\bullet} = \sum_{j,l} Y_{ijl}$, $Y_{\bullet j\bullet} = \sum_{i,l} Y_{ijl}$, $Y_{\bullet\bullet\bullet} = \sum_{i,j,l} Y_{ijl}$.

The Anova Table 3.3 and Type III Sum of squares Table 3.4 can be easily obtained.

Tukey (1949) developed an ingenious method in which the interaction between the ith block and jth treatment is taken as $\gamma \beta_i \tau_j$, so that the parameter γ accounts for all interaction terms. Returning back to the notation and terminology of Sec. 3.1, the sum of squares for non-additivity is calculated as

$$SS_{NA} = \frac{\left\{ \sum_{i,j} Y_{ij} \hat{\beta}_i \hat{\tau}_j \right\}^2}{\left(\sum_i \hat{\beta}_i^2 \right) \left(\sum_j \hat{\tau}_j^2 \right)}.$$

The interaction is tested using an F-statistic

$$F = \frac{\text{SS}_{\text{NA}}}{\left(R_0^2 - \text{SS}_{\text{NA}}\right)/((v-1)(b-1)-1)},$$

with 1 and $(v-1)(b-1)-1$ degrees of freedom and R_0^2 is given by (3.7). For further details justifying the F-distribution, see Scheffè (1959) or Tukey (1949).

4

Balanced Incomplete Block Designs — Analysis and Combinatorics

4.1 Definitions and Basic Results

Balanced Incomplete Block (BIB) designs pose challenging problems in construction, non-existence and combinatorial properties. They have a wide range of applications for diversified problems not originally intended for them. They are variance balanced, combinatorially balanced and efficiency balanced as discussed in Chap. 2. In this chapter we will study their analysis, combinatorics and will discuss the applications in the next chapter.

We formally define a BIB design in Definition 4.1.

Definition 4.1 A *BIB design* is an arrangement of v symbols in b sets each of size $k(<v)$, such that

1. every symbol occurs atmost once in a set
2. every symbol occurs in r sets
3. every pair of distinct symbols occurs together in λ sets.

In experimental settings the symbols are treatments and sets are blocks. In the following arrangement of 7 symbols in 7 sets,

$$(0, 1, 3); (1, 2, 4); (2, 3, 5); (3, 4, 6); (4, 5, 0); (5, 6, 1); (6, 0, 2); \qquad (4.1)$$

the set size is 3; every symbol occurs in 3 sets; and every pair of distinct symbols occurs together in 1 set. Hence it is a BIB design with $v = 7 = b, r = 3 = k, \lambda = 1$. In the following arrangement of 9 symbols in 12 sets,

$$(0, 1, 2); (3, 4, 5); (6, 7, 8); (0, 3, 6); (1, 4, 7); (2, 5, 8);$$
$$(0, 4, 8); (1, 5, 6); (2, 3, 7); (0, 5, 7); (1, 3, 8); (2, 4, 6); \qquad (4.2)$$

the set size is 3; every symbol occurs in 4 sets and every pair of distinct symbols occurs together in 1 set. Hence it is a BIB design with $v = 9$, $b = 12$, $r = 4$, $k = 3$, $\lambda = 1$.

v, b, r, k and λ are known as the parameters of the BIB design. Counting the number of units used in terms of symbols as well as sets, we get

$$vr = bk. \tag{4.3}$$

Taking all the r sets where a given symbol θ occurs, we can form $r(k-1)$ pairs of symbols with θ, and from the definition these pairs must be all pairs of other $v-1$ symbols, with θ, each pair occurring in λ sets. Hence

$$r(k-1) = \lambda(v-1). \tag{4.4}$$

By writing the incidence matrix $N = (n_{ij})$ where $n_{ij} = 1(0)$ according as the ith symbol occurs (does not occur) in the jth set, we have

$$NN' = (r - \lambda)I_v + \lambda J_v, \tag{4.5}$$

with

$$|NN'| = rk(r - \lambda)^{v-1}, \tag{4.6}$$

which is non-singular. Hence

$$v = \text{Rank } (NN') = \text{Rank } (N) \leq b. \tag{4.7}$$

This inequality is originally due to Fisher (1940) and the incidence matrix argument is due to Bose (1949).

By taking all possible combinations of k symbols from v symbols, we get a BIB design with parameters

$$v, b = \binom{v}{k}, \quad r = \binom{v-1}{k-1}, k, \quad \lambda = \binom{v-2}{k-2}, \tag{4.8}$$

and this design is called an irreducible BIB design.

A BIB design is said to be symmetric if $v = b$ and consequently $r = k$; otherwise, asymmetric. For a symmetric BIB design, the incidence matrix N is non-singular, and N commutes with J_v. Pre-multiplying by N' and post-multiplying by $(N')^{-1}$, from (4.5) we get

$$N'N = (r - \lambda)I_v + \lambda J_v, \tag{4.9}$$

for a symmetric BIB design. This implies that every distinct pair of sets of a symmetric BIB design has λ common symbols.

Sometimes the sets of a plan of a BIB design are not all distinct and some sets are repeated. For the plan of a BIB design with parameters $v = 7, b = 35, r = 15,$ $k = 3, \lambda = 5$, one can take 5 copies of plan (4.1), or all possible combinations of 3 symbols selected from the 7 symbols. These two solutions are structurally different and hence one cannot get one solution from the other by permuting the symbols and/or sets. In that sense, they are called non-isomorphic. We introduce some terminology in this connection in Definition 4.2.

Definition 4.2 Given a plan D of a BIB design with parameters v, b, r, k, λ having incidence matrix N, we define the support of the design, D^*, to be the class of distinct sets in D. The number of sets in D^*, denoted by b^*, is called the support size of D. Let the ith set in D^* be repeated f_i times in D for $i = 1, 2, \ldots, b^*$. Then $f' = (f_1, f_2, \ldots, f_{b^*})$ is the frequency vector.

The intrablock analysis and analysis with recovery of interblock information is unaffected whether the design contains repeated sets or not and hence the statistical analysis and optimality of these designs are unaffected whether the design contains repeated blocks or not. However, Raghavarao, Federer and Schwager (1986) considered a linear model useful in market research and intercropping experiments that will distinguish designs with different support sizes and we will discuss these results in Sec. 4.7.

Van Lint (1973) observed that most solutions given by Hanani (1961) have repeated sets, and he wondered whether Hanani could have shown his results had repeated sets been disallowed in the solutions of the designs. No BIB design exists with repeated sets having parameters $v = 2x + 2, b = 4x + 2, k = x + 1$ was demonstrated by Parker (1963), when x is even, and by Seiden (1963), when x is odd. Van Buggenhaut (1974) showed that BIB designs with parameters $v \not\equiv 2$ (mod 3), $k = 3, \lambda = 2$ always exist without repeated sets. Foody and Hedayat (1977), Van Lint (1973), and Van Lint and Ryser (1972) studied the BIB designs with repeated sets in some detail.

4.2 Intra- and Inter-Block Analysis

Using the same notation as in Chap. 2, in the case of a BIB design, we have

$$\begin{aligned} C_{\tau|\beta} &= rI_v - \frac{1}{k}\{(r - \lambda)I_v + \lambda J_v\} \\ &= \frac{\lambda v}{k}I_v - \frac{\lambda}{k}J_v, \end{aligned} \tag{4.10}$$

with its g-inverse

$$C_{\tau|\beta}^- = \frac{k}{\lambda v}I_v.$$

Hence a solution of the reduced normal equations estimating $\boldsymbol{\tau}$, eliminating $\boldsymbol{\beta}$ is

$$\hat{\boldsymbol{\tau}} = \frac{k}{\lambda v}\mathbf{Q}_{\tau|\beta}. \tag{4.11}$$

For a symmetric BIB design, we have $C_{\beta|\tau} = C_{\tau|\beta}$. The following can be easily verified and we can set the ANOVA Table 4.1 and Type III Sum of Squares Table 4.2.
Here

$$SS_T = \sum_{i,j} Y_{ij}^2 - \frac{Y_{\bullet\bullet}^2}{vr}, \qquad SS_{Tr} = \sum_{\ell} \frac{T_{\ell}^2}{r} - \frac{Y_{\bullet\bullet}^2}{vr},$$

$$SS_B = \sum_{i} \frac{B_i^2}{k} - \frac{Y_{\bullet\bullet}^2}{vr}, \qquad SS_{Tr|B} = \frac{k}{\lambda v}\,\mathbf{Q}'_{\tau|\beta}\mathbf{Q}_{\tau|\beta},$$

$$SS_{B|Tr} = SS_B + SS_{Tr|B} - SS_{Tr}, \qquad R_0^2 = SS_T - SS_B - SS_{Tr|B}.$$

The equality of treatment effects can be tested by comparing p_1 with the chosen level α and the equality block effects can be tested by comparing p_2 with α.

The estimated variance of any estimated elementary contrast of treatment effects is $(2k/\lambda v)\,\hat{\sigma}^2$ and this can be used to test the significance or setting confidence intervals of elementary contrasts of treatment effects.

Now let us assume that the block effects are random and σ_b^2 is the population variance of block effects. Then from Chap. 2, we know that $\hat{\sigma}_b^2$ is estimated by

$$\hat{\sigma}_b^2 = \frac{MS_{B|Tr} - \hat{\sigma}^2}{(bk - v)/(b - 1)}. \tag{4.12}$$

Table 4.1. ANOVA.

Source	df	SS	MS
Model	$v + b - 2$	$SS_B + SS_{Tr|B}$	
Error	$vr - v - b + 1$	R_0^2	$\hat{\sigma}^2$
	$(=v)$		

Table 4.2. Type III sum of squares.

Source	df	SS	MS	F	p
Treatments	$v - 1$	$(k/\lambda v)\mathbf{Q}'_{\tau|\beta}\mathbf{Q}_{\tau|\beta}$	$MS_{Tr|B}$	$MS_{Tr|B}/\hat{\sigma}^2$	p_1
Blocks	$b - 1$	$SS_{B|Tr}$	$MS_{B|Tr}$	$MS_{B|Tr}/\hat{\sigma}^2$	p_2

$p_1 = P(F(v - 1, v) \geq F_{Tr|B}(\text{Cal}))$, $p_2 = (F(b - 1, v) \geq F_{B|Tr}(\text{cal}))$ and the right side of the inequality is the calculated F-statistic.

Let $\hat{w} = 1/\hat{\sigma}^2$, $\hat{w}' = 1/\left(\hat{\sigma}^2 + k\hat{\sigma}_b^2\right)$. Then the estimated τ, given by $\tilde{\tau}$ recovering the interblock information, is

$$\tilde{\tau} = \left(\hat{w}C_{\tau|\beta} + \frac{\hat{w}'}{k}NN'\right)^{-1}\left(\hat{w}Q_{\tau|\beta} + \frac{\hat{w}'}{k}NB\right)$$

$$= \frac{k}{\hat{w}\lambda v + \hat{w}'(r - \lambda)}\left\{I_v + \frac{\lambda(\hat{w} - \hat{w}')}{\hat{w}'rk}J_v\right\}\left\{\hat{w}T - \frac{(\hat{w} - \hat{w}')}{k}NB\right\}. \quad (4.13)$$

The equality of treatment effects is tested using the test statistic $\tilde{\tau}'\left(\hat{w}T - \frac{(\hat{w}-\hat{w}')}{k}NB\right)$ which has an approximate $\chi^2(v - 1)$ distribution. The estimated dispersion matrix of $\tilde{\tau}$ is

$$\frac{k}{\hat{w}\lambda v + \hat{w}'(r - \lambda)}\left\{I_v + \frac{\lambda(\hat{w} - \hat{w}')}{\hat{w}'rk}J_v\right\}$$

and it can be used to test hypothesis and/or set confidence intervals for the contrasts of treatment effects.

For a given set of parameters v, b, r, k, where $k < v$, BIB designs are universally optimal, as discussed in Sec. 2.10, whenever they exist, and this result follows from the theorem of that section.

4.3 Set Structures and Parametric Relations

If N_0 is the incidence matrix of any t sets of the solution of a BIB design with parameters v, b, r, k, λ, then

$$S_t = N_0'N_0$$

is called the structural matrix of the t chosen sets and

$$C_t = \lambda k J_t + r(r - \lambda)I_t - rS_t \quad (4.14)$$

is called the characteristic matrix of the t chosen sets.

Connor (1952) proved the following two theorems.

Theorem 4.1 *If C_t is the characteristic matrix of any t chosen sets of a BIB design with parameter v, b, r, k and λ, then*

1. $|C_t| \geq 0$, *if* $t < b - v$
2. $|C_t| = 0$, *if* $t > b - v$
3. $kr^{-t+1}(r - \lambda)^{(v-t-1)}|C_t|$ *is a perfect square if* $t = b - v$.

If $t = 2$, from Theorem 4.1, we get

Theorem 4.2 *For any BIB design, the number of common symbols, s_{ij}, between the ith and jth sets of the design satisfy the inequality*

$$-(r - \lambda - k) \le s_{ij} \le \frac{2\lambda k + r(r - \lambda - k)}{r}. \tag{4.15}$$

Theorem 4.1 can be used to see whether the sets

$$\begin{aligned} &(0, 1, 2, 3, 4, 5) \\ &(6, 7, 8, 9, 10, 11) \\ &(0, 1, 2, 6, 7, 8) \\ &(0, 1, 3, 6, 7, 9) \end{aligned} \tag{4.16}$$

can be part of the completed solution of the BIB design with parameters $v = 16$, $b = 24, r = 9, k = 6, \lambda = 3$. The characteristic matrix in this case is

$$C_4 = \begin{pmatrix} 28 & 18 & -9 & -9 \\ 18 & 28 & -9 & -9 \\ -9 & -9 & 28 & -18 \\ -9 & -9 & -18 & 28 \end{pmatrix}$$

with a positive determinant. Hence the sets (4.16) can be part of the completed solution.

Theorems 4.1 and 4.2 along with bounds on common symbols between any t sets are also discussed in Raghavarao (1971).

S. M. Shah (1975a) obtained upper and lower bounds for the number of sets between which no symbol is common and an upper bound for the number of sets not containing m given symbols. Bush (1977a) pointed out that the lower bound in Shah's first problem is vacuous and improved the bound for the second problem. These results are given in the next two theorems.

Theorem 4.3 *If d denotes the number of sets in a BIB design with parameters v, b, r, k, λ which are pairwise disjoint, then d satisfies*

$$\max(\theta_1, 1) \le d \le \min(\theta_2, v/k), \tag{4.17}$$

where $\theta_1 < \theta_2$ are the roots of the quadratic equation

$$\lambda k d^2 + (r^2 - 2rk - r\lambda + r - \lambda)d - (r - k - \lambda)b = 0.$$

Proof. The d mutually disjoint sets have dk symbols and let the jth set of the remaining $(b - d)$ sets contain ℓ_j of those dk symbols. Then

$$\sum_{j=1}^{b-d} \ell_j = dk(r - 1), \quad \sum_{j=1}^{b-d} \ell_j(\ell_j - 1) = dk(dk\lambda - \lambda - k + 1),$$

and it follows that

$$\sum_{j=1}^{b-d} \ell_j^2 = dk(dk\lambda - \lambda - k + r).$$

From Cauchy–Schwartz's inequality we get

$$\lambda k d^2 + (r^2 - 2rk - r\lambda + r - \lambda)d - (r - k - \lambda)b \leq 0,$$

and hence

$$\theta_1 \leq d \leq \theta_2. \tag{4.18}$$

Obviously

$$1 \leq d \leq v/k. \tag{4.19}$$

Combining (4.18) and (4.19), we get the required result (4.17).

The following lemma given in Bush (1977a) will be needed to prove the next theorem.

Lemma 4.1   *(Schur's Lemma) If $\sum j\ell_j = A$ and $\sum j^2\ell_j = B$ and $(A \leq B)$, then $\sum \ell_j$ is minimum if $\ell_j = 0$ except when $j = p - 1$ or p, where p is the smallest integer such that $B \leq pA$.*

We then have

Theorem 4.4   *The number of sets d^* containing none of the m specified symbols is given by the inequality*

$$d^* \leq b - m[2r(2p - 1) - m\lambda + \lambda]/[p(p - 1)], \tag{4.20}$$

where p is the smallest integer such that $p \geq (m\lambda - \lambda + r)/r$.

Proof. Let ℓ_j be the number of sets in which j of the specified m symbols occur among the $b - d^*$ sets. Then

$$\sum_{j=1}^{\min(k,m)} j\ell_j = mr, \quad \sum_{j=1}^{\min(k,m)} j(j - 1)\ell_j = m(m - 1)\lambda, \tag{4.21}$$

from which we get

$$\sum_{j=1}^{\min(k,m)} j^2 \ell_j = m[(m-1)\lambda + r].\tag{4.22}$$

From Lemman 4.1 noting $A = mr$, $B = m[(m-1)\lambda + r]$, p should be selected such that $p \geq B/A = (m\lambda - \lambda + r)/r$. Then, ℓ_{p-1} and ℓ_p can be solved from (4.21) and (4.22) to be

$$\ell_{p-1} = m[pr - (m-1)\lambda - r]/(p-1),$$
$$\ell_p = m[(m-1)\lambda - pr + 2r]/p,$$

and $\min \sum \ell_j = m[2r(2p-1) - m\lambda + \lambda]/[p(p-1)]$. Thus

$$b - d^* \geq m[2r(2p-1) - m\lambda + \lambda]/[p(p-1)],$$

from which the inequality (4.20) follows.

Let $D(v)$ denote the maximum number of pairwise disjoint triples in Steiner's triple systems, which are BIB designs with $k = 3$ (see Raghavarao, 1971). Teirlinck (1973) showed that $D(3v) \geq 2v + D(v)$ for all $v \geq 3$ and $v \equiv 1$ or 3 (mod 6). Rosa (1975) showed that $D(2v + 1) \geq v + 1 + D(v)$.

Parker (1975) using Connor's characteristic matrix ideas showed that in a Steiner's triple system with $v = 19$, each disjoint pair of sets is contained in at least 43 quadruples of pairwise disjoint sets. Similarly in a Steiner system with $v = 25$, each disjoint pair of sets is contained in a pairwise quintuple of sets.

Shrikhande and Raghavarao (1963) introduced α-resolvability and affine α-resolvability for a general block design, which in the context of BIB designs can be defined as follows:

Definition 4.3 A BIB design with parameters v, b, r, k, λ is said to be α-resolvable if the b sets can be grouped into t classes of β sets such that in each class of β sets every symbol is replicated α times. An α-resolvable BIB design is called *affine α-resolvable* if every pair of sets of the same class intersect in $k - r + \lambda$ symbols and every pair of sets from different classes intersect in k^2/v symbols.

Obviously

$$v\alpha = k\beta, \quad b = t\beta, \quad r = t\alpha.\tag{4.23}$$

When $\alpha = 1$, we get resolvable and affine resolvable BIB designs as defined by Bose (1942).

Shrikhande and Raghavarao (1963) proved the following theorem:

Theorem 4.5 *For an α-resolvable BIB design,*

$$b \geq v + t - 1$$

with equality holding if and only if the design is affine α-resolvable.

Hughes and Piper (1976) also obtained the above result if each symbol occurs α_i times in the ith class for $i = 1, 2, \ldots, t$. Ionin and Shrikhande (1998) extended the result of Theorem 4.5 to combinatorially balanced designs with unequal set sizes.

It is known (see Raghavarao, 1971) that the parameters of an affine resolvable BIB design can be expressed in terms of 2 integral parameters n, t as

$$v = n^2[(n-1)t + 1], \quad b = n(n^2 t + n + 1), \quad r = n^2 t + n + 1,$$
$$k = n[(n-1)t + 1], \quad \lambda = nt + 1. \tag{4.24}$$

Generalizing the above, Kageyama (1973b) showed that the parameters of affine α-resolvable BIB design can be expressed in terms of three integral parameters $\alpha(\geq 1)$, $\beta(\geq 2)$, and j $(\geq(1-\alpha)/\beta_1)$ as follows:

$$v = \frac{\beta}{\alpha}[\beta_1(\beta - 1)j + \beta\alpha], \quad b = \frac{\beta}{\alpha}[\beta\beta_1 j + (\beta + 1)\alpha],$$
$$r = \beta\beta_1 j + (\beta + 1)\alpha, \qquad k = \beta_1(\beta - 1)j + \beta\alpha, \tag{4.25}$$
$$\lambda = \beta_1\alpha j + \alpha^2 + \alpha(\alpha - 1)/(\beta - 1),$$

where $\beta_1 = \beta/g$, g being the greatest common divisor of α and β. When $g = 1$, take $j \geq 0$. In (4.25), taking $\alpha = 1$, $\beta = \beta_1 = n$, $j = t$, we get (4.24).

We will now derive some inequalities on the parameters.

Theorem 4.6 *(Das (1954), Kageyama and Ishii (1975))* $k(v-1)/r$ *is an integer in a BIB design with parameters* v, b, r, k, λ *if and only if the greatest common divisor of* b, r, λ, *denoted by* (b, r, λ) *is 1.*

Proof. Let $(b, r, \lambda) = 1$, and let $(b, \lambda) = c$. Clearly $(c, r) = 1$. Further, let $\lambda = c\lambda_1$. From $\lambda(v-1) = r(k-1)$, it follows that $c|(k-1)$, and $\lambda_1|r(k-1)$. Let $\lambda_1 = \lambda_2\lambda_3$, where we allow λ_2 or λ_3 to be 1, and without loss of generality we assume $\lambda_2|r, \lambda_3|(k-1)$. Since $vr = bk$ and as $\lambda_2|b$ (otherwise, $(b, \lambda) > c$), we have $\lambda_2|k$. Hence $\lambda_2\lambda_3 = \lambda_1|(k-1)k$ and hence $c\lambda_1 = \lambda|(k-1)k$. Thus $k(k-1)/\lambda$ is an integer. From the basic relations we have $k(k-1)/\lambda = k(v-1)/r$. Thus $k(v-1)/r$ is an integer.

When $(b, r, \lambda) \neq 1$, we have $k(v-1)/r$ not necessarily integral as can be seen from the parameters $v = 22, b = 44, r = 14, k = 7, \lambda = 4$ for which the design exists (see series 71 of Table 4.4). This completes the proof.

For resolvable BIB designs, it is known from Theorem 4.5 that $b \geq v + r - 1$ and the resolvable design becomes affine resolvable if $b = v + r - 1$. Kageyama (1971) improved this result to the following:

Theorem 4.7 *In a BIB design with parameters* $v = nk, b = nr, r, k, \lambda,$ *if $b > v + r - 1$, then*

$$b \geq 2v + r - 2. \tag{4.26}$$

Proof. Since $b = vr/k$, it follows that

$$b - (v + r - 1) = (v - 1)(r - \lambda - k)/k.$$

Now $(k, v-1) = 1$ as $v = nk$. Since $b-(v+r-1)$ is positive integral, $(r-\lambda-k)/k$ is a positive integer, say, t. Then we have $b = (t+1)(v-1)+r$ and (4.26) follows.

Since $b > v + r - 1$ for a resolvable BIB design which is not affine resolvable, we have

Corollary 4.7.1 *For a resolvable BIB design which is not affine resolvable, the inequality* (4.26) *holds.*

Generalizing the above corollary to α-resolvable designs, Kageyama (1973a) obtained the following theorem:

Theorem 4.8 *For an α-resolvable BIB design with parameters* $v, b = t\beta,$ $r = t\alpha, k, \lambda,$ *which is not affine α-resolvable, we have*

$$b \geq \frac{2(v - 1)}{\alpha} + r. \tag{4.27}$$

The following lemma due to Ryser will be required to prove the main result of Theorem 4.9 related to the set structure of BIB design with repeated sets.

Lemma 4.2 *Let X and Y be $n \times n$ real matrices such that*

$$XY = D + \mathbf{uu}',$$

where $D = \mathrm{diag}(r_1 - \lambda_1, r_2 - \lambda_2, \ldots, r_n - \lambda_n)$, $\mathbf{u}' = \left(\sqrt{u_1}, \sqrt{u_2}, \ldots, \sqrt{u_n}\right)$ and the scalars $r_i - \lambda_i$ and u_j are positive and non-negative, respectively. Then

$$YD^{-1}X = I_n + \{1 + \mathbf{u}'D^{-1}\mathbf{u}\}^{-1}YD^{-1}\mathbf{uu}'D^{-1}X. \tag{4.28}$$

Proof. Due to the conditions on $r_i - \lambda_i$ and u_j, we note that XY is positive definite and hence XY, X and Y are non-singular. Hence

$$(XY)^{-1} = Y^{-1}X^{-1} = D^{-1} - \{1 + \mathbf{u}'D^{-1}\mathbf{u}\}^{-1}D^{-1}\mathbf{uu}'D^{-1},$$

from which we get

$$D^{-1} = Y^{-1}X^{-1} + \{1 + \mathbf{u}'D^{-1}\mathbf{u}\}^{-1}D^{-1}\mathbf{u}\mathbf{u}'D^{-1}.$$

Pre-multiplying and post-multiplying by Y and X respectively we get (4.28).

Theorem 4.9 *With the notation of Definition 4.2, we have*

$$N^{*\prime}N^* = (r - \lambda)F^{-1} + \frac{\lambda k}{r}J_{b^*} - W, \tag{4.29}$$

where W is a positive semi-definite matrix of order b^*, of rank $b^* - v$, $F = D(f_1, f_2, \ldots, f_{b^*})$, and N^* is the incidence matrix of the support of the BIB design.

Proof. Since $NN' = (r - \lambda)I_v + \lambda J_v$, it follows that

$$N^*FN^{*\prime} = (r - \lambda)I_v + \lambda J_v.$$

Thus the $v \times b^*$ matrix $N^*F^{1/2}$ is of rank v, and

$$b^* \geq v.$$

Put

$$X' = [F^{1/2}N^{*\prime}|A'],$$

where A is a $(b^* - v) \times b^*$ matrix satisfying

$$AA' = (r - \lambda)I_{b^*-v}, \quad AF^{1/2}N^{*\prime} = O_{b^*-v,v}.$$

Now

$$XX' = \begin{pmatrix} (r - \lambda)I_v + \lambda J_v & O_{v,b^*-v} \\ O_{b^*-v,v} & (r - \lambda)I_{b^*-v} \end{pmatrix}$$

and using Lemma 4.2 with $Y = X'$, $n = b^*$, $r_1 = r_2 = \cdots = r_{b^*} = r$, $\lambda_1 = \lambda_2 = \cdots = \lambda_{b^*} = \lambda$, $u_1 = u_2 = \cdots = u_v = \sqrt{\lambda}$, $u_{v+1} = \cdots = u_{b^*} = 0$, and straight forward simplification we get the following two expressions for $X'X$

$$X'X = (r - \lambda)I_{b^*} + \frac{\lambda k}{r}F^{1/2}J_{b^*}F^{1/2}$$

$$= F^{1/2}N^{*\prime}N^*F^{1/2} + A'A.$$

Equating the above two expressions, we have

$$N^{*\prime}N^* = (r - \lambda)F^{-1} + \frac{\lambda k}{r} J_{b^*} - F^{-1/2} A'AF^{-1/2}.$$

Putting $W = F^{-1/2} A'AF^{-1/2}$ we get (4.29) and it can be easily noted that W is a positive semi-definite matrix of order b^* and of rank $b^* - v$.

Since $W = (w_{ij})$ is a positive semi-definite matrix, $w_{ii} \geq 0$ and $w_{ii}w_{jj} - w_{ij}^2 \geq 0$ for all i and j $(i \neq j)$. This implies the following corollary, making use of Eq. (4.29)

Corollary 4.9.1 *If s_{ij} is the number of common symbols between the ith and jth sets of the support D^* of a BIB design D, we have*

$$f_i \leq b/v, \qquad\qquad i = 1, 2, \ldots, b^* \qquad\qquad (4.30)$$

$$\left(\frac{r}{f_i} - k\right)\left(\frac{r}{f_j} - k\right) \geq \left(\frac{\lambda k - rs_{ij}}{r - \lambda}\right)^2, \quad i, j = 1, 2, \ldots, b^*, i \neq j. \quad (4.31)$$

Mann (1952) obtained (4.30) and (4.31) is an improvement over Connor's inequality given in Theorem 4.2, which can be expressed as

$$(r - k)^2 \geq \left(\frac{\lambda k - rs_{ij}}{r - \lambda}\right)^2. \qquad\qquad (4.32)$$

Van Lint and Ryser (1972) proved the following theorem:

Theorem 4.10 *The support size b^* of a BIB design satisfies $b^* = v$, or $b^* > v+1$. In case $b^* = v$ holds, the BIB design D is obtained by replicating the support, that is $f_1 = f_2 = \cdots = f_{b^*}$.*

4.4 Related Designs

In Sec 4.1 we defined a symmetric BIB design with $v = b$ and any two distinct sets of the design have exactly λ symbols in common. From a symmetric BIB design we can construct a residual design by deleting a set and the symbols of that set from the other sets. By deleting a set and retaining only the symbols of that set in other sets, we can construct a derived design from a symmetric BIB design. The parameters of the residual and derived designs that can be constructed from a symmetric BIB design with parameters $v = b, r = k, \lambda$ are respectively

$$v_1 = v - k, \quad b_1 = v - 1, \quad r_1 = k, \quad k_1 = k - \lambda, \quad \lambda_1 = \lambda \qquad (4.33)$$

and

$$v_2 = k, \quad b_2 = v - 1, \quad r_2 = k - 1, \quad k_2 = \lambda, \quad \lambda_2 = \lambda - 1. \qquad (4.34)$$

The complement of any BIB design is obtained by replacing every set of symbols by its complementary set. The complement of a BIB design with parameters v, b, r, k, λ is a BIB design with parameters

$$v_3 = v, \quad b_3 = v, \quad r_3 = b - r, \quad k_3 = v - k, \quad \lambda_3 = b - 2r + \lambda, \qquad (4.35)$$

with incidence matrix $\bar{N} = J_{v,b} - N$, where N is the incidence matrix of the original design.

A Hadamard matrix H of order n is an $n \times n$ matrix of elements 1 or -1 such that $HH' = nI_n$. A necessary condition for its existence is $n = 2$ or $n \equiv 0 \pmod 4$ and the methods of construction of Hadamard matrices are given in Raghavarao (1971). Without loss of generality by multiplying the columns and rows of H by -1, we can write H when $n = 4t$ in the form

$$H = \begin{pmatrix} 1 & \mathbf{1}'_{4t-1} \\ \mathbf{1}_{4t-1} & H_1 \end{pmatrix}.$$

In H_1 replacing -1 by 0, we get the incidence matrix of a symmetric BIB design with parameters

$$v = 4t - 1 = b, \quad r = 2t - 1 = k, \quad \lambda = t - 1. \qquad (4.36)$$

A BIB design with parameters (4.36) and its complement are both called Hadamard designs. By reversing the process from a solution of BIB design with parameters (4.36), we can construct a Hadamard matrix of order $4t$.

A Hadamard matrix of order 12 is

$$H_{12} = \begin{pmatrix}
1 & 1 & 1 & 1 & 1 & 1 & 1 & 1 & 1 & 1 & 1 & 1 \\
1 & -1 & -1 & 1 & -1 & 1 & -1 & 1 & -1 & 1 & -1 & 1 \\
1 & -1 & 1 & -1 & 1 & 1 & -1 & -1 & -1 & -1 & 1 & 1 \\
1 & 1 & -1 & -1 & 1 & -1 & -1 & 1 & -1 & 1 & 1 & -1 \\
1 & -1 & 1 & 1 & 1 & -1 & 1 & 1 & -1 & -1 & -1 & -1 \\
1 & 1 & 1 & -1 & -1 & -1 & 1 & -1 & -1 & 1 & -1 & 1 \\
1 & -1 & -1 & -1 & 1 & 1 & 1 & -1 & 1 & 1 & -1 & -1 \\
1 & 1 & -1 & 1 & 1 & -1 & -1 & -1 & 1 & -1 & -1 & 1 \\
1 & -1 & -1 & -1 & -1 & -1 & 1 & 1 & 1 & -1 & 1 & 1 \\
1 & 1 & -1 & 1 & -1 & 1 & 1 & -1 & -1 & -1 & 1 & -1 \\
1 & -1 & 1 & 1 & -1 & -1 & -1 & -1 & 1 & 1 & 1 & -1 \\
1 & 1 & 1 & -1 & -1 & 1 & -1 & 1 & 1 & -1 & -1 & -1
\end{pmatrix}.$$

By changing -1 to 0 in H_1 we get the incidence matrix of a BIB design with parameters $v = 11 = b, r = 5 = k, \lambda = 2$, from which we can write the solution in symbols $0, 1, \ldots, 10$ as

$(2, 4, 6, 8, 10)$; $(1, 3, 4, 9, 10)$; $(0, 3, 6, 8, 9)$; $(1, 2, 3, 5, 6)$; $(0, 1, 5, 8, 10)$; $(3, 4, 5, 7, 8)$; $(0, 2, 3, 7, 10)$; $(5, 6, 7, 9, 10)$; $(0, 2, 4, 5, 9)$; $(1, 2, 7, 8, 9)$; $(0, 1, 4, 6, 7)$;

and its complement

$(0, 1, 3, 5, 7, 9)$; $(0, 2, 5, 6, 7, 8)$; $(1, 2, 4, 5, 7, 10)$; $(0, 4, 7, 8, 9, 10)$; $(2, 3, 4, 6, 7, 9)$; $(0, 1, 2, 6, 9, 10)$; $(1, 4, 5, 6, 8, 9)$; $(0, 1, 2, 3, 4, 8)$; $(1, 3, 6, 7, 8, 10)$; $(0, 3, 4, 5, 6, 10)$; $(2, 3, 5, 8, 9, 10)$;

which is also a BIB design with parameters $v = 11 = b, r = 6 = k, \lambda = 3$.

A Latin square of order s is an $s \times s$ square array in s symbols such that each symbol occurs exactly once in each row and each column. A pair of Latin squares of order s is said to be orthogonal if on superimposition every ordered pair of symbols occurs together exactly once. The maximum number of Latin squares of order s that are pairwise orthogonal, is $s - 1$ and they exist when s is a prime or prime power. Such a set of $s - 1$ Latin squares, which are mutually orthogonal is called a complete set of Mutually Orthogonal Latin Squares (MOLS) and their construction is given in Raghavarao (1971).

Let the number of symbols in a BIB design be $v = s^2$, where s is a prime or prime power. Arrange the symbols in an $s \times s$ square array. Form one replication of s sets by writing each set with the symbols occurring in the same row. Form second replication of sets by writing each set with the symbols occurring in the same column. Using each Latin square of the complete set of MOLS, form a replicate of sets by forming the sets with the symbols of the square array that occur in the same position with each of the symbols of the Latin square. From the properties of orthogonal Latin squares the design obtained is an affine resolvable BIB design with parameters

$$v = s^2, \quad b = s(s+1), \quad r = s+1, \quad k = s, \quad \lambda = 1. \qquad (4.37)$$

The solution of the BIB design with parameters (4.37) is a finite affine plane (see Veblen and Bussey, 1906)

Consider $v = 16$ and the symbols are written in a 4×4 array as follows:

$$\begin{array}{cccc}
0 & 1 & 2 & 3 \\
4 & 5 & 6 & 7 \\
8 & 9 & 10 & 11 \\
12 & 13 & 14 & 15
\end{array}$$

The complete set of MOLS of order 4 are

$$
\begin{array}{cccc}
A & B & C & D \\
B & A & D & C \\
C & D & A & B \\
D & C & B & A
\end{array}
$$

$$
\begin{array}{cccc}
A & C & D & B \\
B & D & C & A \\
C & A & B & D \\
D & B & A & C
\end{array}
$$

$$
\begin{array}{cccc}
A & D & B & C \\
B & C & A & D \\
C & B & D & A \\
D & A & C & B
\end{array}
$$

The solution of the BIB design with parameters $v = 16, b = 20, r = 5, k = 4,$ $\lambda = 1$ as discussed here is

$$
\begin{array}{llll}
(0, 1, 2, 3); & (4, 5, 6, 7); & (8, 9, 10, 11); & (12, 13, 14, 15); \\
(0, 4, 8, 12); & (1, 5, 9, 13); & (2, 6, 10, 14); & (3, 7, 11, 15); \\
(0, 5, 10, 15); & (1, 4, 11, 14); & (2, 7, 8, 13); & (3, 6, 9, 12); \\
(0, 7, 9, 14); & (3, 4, 10, 13); & (1, 6, 8, 15); & (2, 5, 11, 12); \\
(0, 6, 11, 13); & (2, 4, 9, 15); & (3, 5, 8, 14); & (1, 7, 10, 12);
\end{array}
\tag{4.38}
$$

which is a finite affine plane in 16 points and 20 lines.

From the solution of the affine resolvable BIB design with parameters (4.37), adding a new symbol i to each set of the replication i for $i = 1, 2, \ldots, s + 1,$ and forming a new set consisting of all the newly added $s + 1$ symbols, we get a symmetric BIB design with parameters

$$
v = s^2 + s + 1 = b, \quad r = s + 1 = k, \quad \lambda = 1.
\tag{4.39}
$$

The solution of the symmetric BIB design with parameters (4.39) is the finite projective plane.

By adding symbols 16, 17, 18, 19, 20 to each of the first 4 sets, second 4 sets, third 4 sets, fourth 4 sets, and fifth 4 sets respectively of (4.38) and adding a new set of symbols $(16, 17, 18, 19, 20)$, we get a symmetric BIB design with parameters

$$v = 21 = b, \quad r = 5 = k, \quad \lambda = 1,$$

which is a finite projective plane in 21 points and lines.

Combining the solutions (or incidence matrices) of BIB designs, we get other BIB designs. The following are some examples.

Theorem 4.11 (*Shrikhande and Raghavarao, 1963*) *Let M be the incidence matrix of a BIB design with parameter $v_1, b_1, r_1, k_1, \lambda_1$ and let $N = [N_1|N_2|\cdots|N_{r_2}]$ be the incidence matrix of a resolvable BIB design with parameters $v_2 = k_2 v_1, b_2 = r_2 v_1, r_2, k_2, \lambda_2$, where N_i is the incidence matrix of the sets of the ith replicate of the resolvable BIB design. Then $M_1 = [N_1 M | N_2 M | \cdots | N_{r_2} M]$ is the incidence matrix of a α-resolvable BIB design with parameters*

$$
\begin{aligned}
v &= v_2, \quad b = b_1 r_2, \quad r = r_1 r_2, \quad k = k_1 k_2, \quad \alpha = r_1, \\
\beta &= b_1, \quad t = r_2, \quad \lambda = r_2 \lambda_1 + \lambda_2 (r_1 - \lambda_1).
\end{aligned}
\tag{4.40}
$$

We prove it by verifying that the columns of $N_i M$ form the ith class of sets, $N_i M \mathbf{1}_{b_1} = r_1 \mathbf{1}_{v_2}$, and

$$M_1 M_1' = (r_1 - \lambda_1)(r_2 - \lambda_1) I_{v_2} + \{r_2 \lambda_1 + \lambda_2 (r_1 - \lambda_1)\} J_{v_2}.$$

Theorem 4.12 (*S.S. Shrikhande, 1962*) *If N_i is the incidence matrix of a BIB design with parameters $v_i, b_i, r_i, k_i, \lambda_i$ satisfying $b_i = 4(r_i - \lambda_i)$ for $i = 1, 2$, then*

$$N = N_1 \otimes N_2 + (J_{v_1, b_1} - N_1) \otimes (J_{v_2, b_2} - N_2),$$

where $\otimes$ denotes the Kronecker product of matrices, is the incidence matrix of the BIB design with parameters

$$
\begin{aligned}
v &= v_1 v_2, \quad b = b_1 b_2, \quad r = r_1 r_2 + (b_1 - r_1)(b_2 - r_2), \\
k &= k_1 k_2 + (v_1 - k_1)(v_2 - k_2), \quad \lambda = r - b/4.
\end{aligned}
\tag{4.41}
$$

Proof is by verification that $N \mathbf{1}_b = r \mathbf{1}_v$, $\mathbf{1}_v' N = k \mathbf{1}_b'$, and

$$NN' = \frac{b}{4} I_v + \left(r - \frac{b}{4} \right) J_v.$$

Theorem 4.13 (*S.S. Shrikhande and Singh, 1962*) *If there exists a BIB design with $\lambda = 1$ and $r = 2k + 1$, then a symmetric BIB design exists with parameters*

$$v^* = 4k^2 - 1 = b^*, \quad r^* = 2k^2 = k^*, \quad \lambda = k^2.
\tag{4.42}$$

Proof. The BIB design with $\lambda = 1$, $r = 2k + 1$, necessarily has the parameters $v = k(2k - 1)$, $b = 4k^2 - 1$, $r = 2k + 1$, k, $\lambda = 1$ and the number of common symbols between any two sets is at most one. Let N be the incidence matrix of such a design. Then

$$N'N = kI_b + M,$$

where M is a matrix of elements 0 or 1. It can be verified that M is the incidence of the BIB design with parameters (4.42).

Let N be the incidence matrix of a symmetric BIB design with parameters $v = b$, $r = k$, λ and let

$$N = \begin{pmatrix} N_1 & N_2 \\ N_3 & N_4 \end{pmatrix},$$

where N_1 is the incidence matrix of a symmetric BIB design with parameters $v_1 = b_1$, $r_1 = k_1$, λ_1. Then the design with incidence matrix N_1 is a subdesign of the design with incidence matrix N.

The symmetric BIB design with parameters $v = 11 = b$, $r = 6 = k$, $\lambda = 3$ given by

$$
\begin{aligned}
&(\underline{0}, \underline{4}, \underline{5}, 6, 8, 9); \ (\underline{1}, \underline{5}, \underline{6}, 7, 9, 10); \ (\underline{2}, \underline{6}, \underline{0}, 8, 10, 7); \ (\underline{3}, \underline{0}, \underline{1}, 9, 7, 8); \\
&(\underline{4}, \underline{1}, \underline{2}, 10, 8, 9); \ (\underline{5}, \underline{2}, \underline{3}, 0, 9, 10); \ (\underline{6}, \underline{3}, \underline{4}, 1, 10, 0); \ (7, 4, 5, 2, 0, 1); \\
&(8, 5, 6, 3, 1, 2); \ (9, 6, 7, 4, 2, 3); \ (10, 7, 8, 5, 3, 4); \hspace{3cm} (4.43)
\end{aligned}
$$

has subdesign, a symmetric BIB design with parameters $v = 7 = b$, $r = 3 = k$, $\lambda = 1$. The subdesign has 7 sets consisting of the underlined symbols in the sets of (4.43).

Given a symmetric BIB design with parameters $v = b$, $r = k$, λ, a symmetric BIB design with parameters $v_0 = b_0$, $r_0 = k_0$, λ which is a subdesign of the first design is called Baer subdesign if $k - \lambda = (k_0 - 1)^2$. For interesting combinatorial results on the subdesigns, we refer to Baartmans and Shrikhande (1981), Bose and Shrikhande (1976), Haemers and Shrikhande (1979), and Kantor (1969).

We close this section with this interesting result of Sprott (1954).

Theorem 4.14 *In the sets of a BIB design and the sets of its complement, every distinct triple of the symbols occur together equally often.*

Proof. Let v, b, r, k, λ be the parameters of the BIB design and $v_3 = v$, $b_3 = b$, $r_3 = b - r$, $k_3 = v - k$, $\lambda_3 = b - 2r + \lambda$ be the parameters of the complement design. Let three distinct symbols θ, ϕ and χ occur together in x sets of the original design. It can be verified that none of θ, ϕ and χ occur in $b - 3r + 3\lambda - x$ sets of the original design. Consequently θ, ϕ and χ occur in $b - 3r + 3\lambda - x$ sets of the complement design and none of them occur in x sets of the complement design. Thus in the original and complement designs altogether the triplet of distinct symbols θ, ϕ and χ occur in $b - 3r + 3\lambda$ sets and this number is independent of the selected triplet of symbols.

4.5 Construction of BIB Designs from Finite Geometries

Let s be a prime or prime power and GF(s) be the Galois field of s elements. For a brief review of Galois fields, see Raghavarao (1971). The points of the finite projective geometry, PG(n, s), are represented by $n+1$ coordinates $(x_0, x_1, \ldots, x_n)$, where $x_i \in$ GF(s), for $i = 0, 1, \ldots, n$ and all x_i are nonzero. The points $(x_0, x_1, \ldots, x_n)$ and $(\rho x_0, \rho x_1, \ldots, \rho x_n)$ for $\rho \in$ GF(s) and $\rho \neq 0$, are the same and hence without loss of generality, we assume the first nonzero coordinate of the point to be 1. Points satisfying $n - m$ linear independent homogeneous equations

$$\sum_{j=0}^{n} a_{ij} x_j = 0, \quad i = 1, 2, \ldots, n - m \tag{4.44}$$

constitute an m-flat of the geometry. The m-flat corresponding to (4.44) is generated by $(m + 1)$ independent points of the geometry. The number of points in the geometry can be verified to be $P_n = (s^{n+1} - 1)/(s - 1)$, and the number of points on an m-flat to be $P_m = (s^{m+1} - 1)/(s - 1)$.

The number of m-flats can be determined by choosing $(m + 1)$ independent points from the geometry and noting that the same flat can be generated by any $(m+1)$ independent points on the given m-flat. Thus the number of m-flats, denoted by $\phi(n, m, s)$, is

$$\phi(n, m, s) = \frac{P_n(P_n - 1)(P_n - P_1) \cdots (P_n - P_{m-1})}{P_m(P_m - 1)(P_m - P_1) \cdots (P_m - P_{m-1})}$$

$$= \frac{(s^{n+1} - 1)(s^n - 1) \cdots (s^{n-m+1} - 1)}{(s^{m+1} - 1)(s^m - 1) \cdots (s - 1)}. \tag{4.45}$$

The number of m-flats going through any given one point (two points) of the geometry is $\phi(n-1, m-1, s)$ ($\phi(n-2, m-2, s)$). Conventionally $\phi(n, m, s) = 0$, if m is negative.

By taking the points of the projective geometry as symbols and m-flats as sets, we get several series of BIB designs (see Bose, 1939; Raghavarao, 1971) with parameters

$$v = P_n, \quad b = \phi(n, m, s), \quad r = \phi(n - 1, m - 1, s),$$
$$k = P_m, \quad \lambda = \phi(n - 2, m - 2, s), \tag{4.46}$$

for positive integral n and m.

From $\mathrm{PG}(n, s)$ by deleting the point $(1, 0, 0, \ldots, 0)$ and all flats going through it, we get finite Euclidean geometry, $\mathrm{EG}(n, s)$. In $\mathrm{EG}(n, s)$ there are $E_n = s^n$ points with coordinates $(x_1, x_2, \ldots, x_n)$, where $x_i \in \mathrm{GF}(s)$. The m-flats satisfy a system of $n - m$ linear, independent nonhomogeneous equations and the number of points on an m-flat is $E_m = s^m$. The number of m-flats is $\phi(n, m, s) - \phi(n - 1, m, s)$. The number of m-flats passing through one point (two points) are $\phi(n - 1, m - 1, s)(\phi(n - 2, m - 2, s))$.

Again, identifying the points of $\mathrm{EG}(n, s)$ as symbols and m-flats as sets, we get BIB designs (see Bose, 1939; Raghavarao, 1971) with parameters

$$v = s^n, \quad b = \phi(n, m, s) - \phi(n - 1, m, s),$$
$$r = \phi(n - 1, m - 1, s), \quad k = s^m, \quad \lambda = \phi(n - 2, m - 2, s). \tag{4.47}$$

Examples of designs constructed by these methods are given in Raghavarao (1971). For further details on projective geometries, the interested reader is referred to Hirschfeld (1979).

4.6 Construction by the Method of Differences

The method of differences originally introduced by Bose (1939) is a powerful tool of constructing BIB designs. By modifying the requirements appropriately, we can use this technique to construct other types of designs. Several series of BIB designs were constructed by this method. We will first introduce some terminology and prove some useful results.

Let M be a module of m elements denoted by $0, 1, \ldots, m - 1$. To each element $u \in M$, we associate t symbols $u_1, u_2, \ldots, u_t$, so that we have $v = mt$ symbols. The symbols $0_i, 1_i, \ldots, (m - 1)_i$ are said to belong to the ith class. Given two symbols u_i and u'_i, the difference $u - u'$ and $u' - u$ taken mod m are known as pure difference, of type (i, i). Given two symbols u_i and u'_j, the difference $u - u'$ taken mod m is known as mixed difference of type (i, j). There are t types of pure differences and $t(t - 1)$ types of mixed differences. Given a set S_α of v symbols, we define the set $S_{\alpha\theta}$ to be the set obtained from S_α by adding θ of the module M to each element in S_α keeping the class unaltered. For example if $M = \{0, 1, 2, 3, 4\}$, $t = 2$

and $S_\alpha = (2_1, 3_2, 4_1)$, then $S_{\alpha 2} = ((2+2)_1, (3+2)_2, (4+2)_1) = (4_1, 0_2, 1_1)$. We now prove

Theorem 4.15 *If we form w sets $S_1, S_2, \ldots, S_w$ each of size k in the v symbols such that*

1. *each of the t classes is represented r times $(tr = wk)$.*
2. *Among the $wk(k-1)$ differences, each nonzero element of M occurs λ times as a pure difference of type (i, i) for $i = 1, 2, \ldots, t$, and each element of M occurs λ times as a mixed difference of type (i, j) for $i \neq j; i, j = 1, 2, \ldots, t$, $[wk(k-1) = \lambda t(m-1) + \lambda t(t-1)m]$,*

then the sets $S_{\alpha\theta}$ for $\alpha = 1, 2, \ldots, w$ and $\theta = 0, 1, 2, \ldots, m-1$ form a BIB design with parameters

$$v = mt, \quad b = wm, \quad r, k, \lambda. \tag{4.48}$$

Proof. Given a symbol x_i, it occurs in the sets $S_{\alpha\theta}$, whenever $u_i \in S_\alpha$, $x = u + \theta \pmod{m}$, and from condition 1 of the theorem x_i occurs in r sets. Given a pair of symbols x_i, y_i with $x - y = d(\neq 0)$, they occur together in sets $S_{\alpha\theta}$ if there exist $u_i, u_i' \in S_\alpha$ and $u - u' = d$. As there are λ pairs u_i and u_i' satisfying $u - u' \equiv d \pmod{m}$, x_i, y_i occur together in λ sets. Similarly the symbols x_i, y_j also occur in λ sets and the theorem is proved.

The sets $S_1, S_2, \ldots, S_w$ are called the initial sets. Often $t = 1$, so that there is only one type of difference.

One can modify the theorem and add an extra symbol ∞ such that

$$\infty + u = \infty, \quad \infty - u = \infty, \quad u - \infty = \infty$$

and modify Theorem 4.15 to construct BIB designs with $v = mt + 1$ symbols (see Bose, 1939; Raghavarao, 1971).

In the following corollaries, we give some applications of Theorem 4.15.

Corollary 4.15.1 *Let $v = 6t + 1$ be a prime or prime power and x be a primitive root of $GF(v)$. Then the t initial sets*

$$S_\alpha = (0, x^\alpha, x^{2t+\alpha}, x^{4t+\alpha}), \quad \alpha = 0, 1, \ldots, t-1,$$

by the method of Theorem 4.15 gives a solution of the BIB design with parameters

$$v = 6t + 1, \quad b = t(6t + 1), \quad r = 4t, \quad k = 4, \quad \lambda = 2. \tag{4.49}$$

Proof. The $12t$ differences formed from S_α, $\alpha = 0, 1, \ldots, t-1$ are

$$\pm x^\alpha, \pm x^{2t+\alpha}, \pm x^{4t+\alpha}, \pm x^\alpha(x^{2t}-1), \pm x^\alpha(x^{4t}-1),$$
$$\pm x^{2t+\alpha}(x^{2t}-1), \qquad \alpha = 0, 1, \ldots, t-1.$$

Here $x^{6t} = 1$, and hence $x^{3t} = -1$, and $x^{2t} + 1 = x^t$. Hence the above differences can be written as

$$x^\alpha, x^{t+\alpha}, x^{2t+\alpha}, x^{3t+\alpha}, x^{4t+\alpha}, x^{5t+\alpha}, (x^{2t}-1)x^\alpha, (x^{2t}-1)x^{t+\alpha},$$
$$(x^{2t}-1)x^{2t+\alpha}, (x^{2t}-1)x^{3t+\alpha}, (x^{2t}-1)x^{4t+\alpha}, (x^{2t}-1)x^{5t+\alpha},$$
$$\alpha = 0, 1, \ldots, t-1.$$

Hence each nonzero element of $GF(v)$ occurs twice among all differences and the corollary is proved.

Corollary 4.15.2 *Let $v = 4t + 3$ be a prime or prime power and x be a primitive root of $GF(v)$. The initial set*

$$S_1 = (x^2, x^4, x^6, \ldots, x^{4t+2})$$

by the method of Theorem 4.15 give a solution of the BIB design with parameters

$$v = 4t + 3 = k, \quad r = 2t + 1 = k, \quad \lambda = t. \tag{4.50}$$

Proof. All the $(2t+1)2t$ differences from S_1 are

$$\pm x^2(x^2-1), \pm x^4(x^2-1), \ldots, \pm x^{4t}(x^2-1), \pm x^2(x^4-1),$$
$$\pm x^4(x^4-1), \ldots, \pm x^{4t-2}(x^4-1), \pm x^2(x^6-1),$$
$$\pm x^4(x^6-1), \ldots, \pm x^{4t-4}(x^6-1), \ldots, \pm x^2(x^{4t}-1).$$

Note that $x^{4t+2} = 1$, $x^{2t+1} = -1$, $\pm(x^{4t-2i}-1) = \pm x^{4t-2i}(x^{2i+2}-1)$. Using these results, the differences become

$$\pm x^2(x^2-1), \pm x^4(x^2-1), \ldots, \pm x^{4t+2}(x^2-1),$$
$$\pm x^2(x^4-1), \pm x^4(x^4-1), \ldots, \pm x^{4t+2}(x^4-1),$$
$$\vdots$$
$$\pm x^2(x^{2t}-1), \pm x^4(x^{2t}-1), \ldots, \pm x^{4t+2}(x^{2t}-1).$$

The elements in each row is a complete replication of the nonzero elements of $GF(v)$ and hence all nonzero elements occur t times among the differences. Thus the corollary is proved.

Corollary 4.15.3 *Let M be a module of $2t + 1$ elements $0, 1, \ldots, 2t$ and to each element u, we associate 3 symbols u_1, u_2, u_3. The $3t + 1$ initial sets*

$$(1_1, \quad 2t_1, \quad 0_2); \quad (2_1, (2t - 1)_1, 0_2); \quad \ldots; \quad (t_1, (t + 1)_1, 0_2);$$
$$(1_2, \quad 2t_2, \quad 0_3); \quad (2_2, (2t - 1)_2, 0_3); \quad \ldots; \quad (t_2, (t + 1)_2, 0_3);$$
$$(1_3, \quad 2t_3, \quad 0_1); \quad (2_3, (2t - 1)_3, 0_1); \quad \ldots; \quad (t_3, (t + 1)_3, 0_1);$$
$$(0_1, \quad 0_2, \quad 0_3)$$

by the method of Theorem 4.15 give a solution of the BIB design with parameters

$$v = 6t + 3, \quad b = (2t + 1)(3t + 1), \quad r = 3t + 1, \quad k = 3, \quad \lambda = 1. \quad (4.51)$$

Proof. The pure differences of (i, i) type for $i = 1, 2, 3$ are $2t - 1, 2, 2t - 3, 4, \ldots, 1, 2t$ and each nonzero element of M occurs once as a pure difference. The mixed differences $(1, 2)$ for example are

$$1, 2t, 2, 2t - 1, \ldots, t, t + 1 \text{ and } 0,$$

and all the elements of M occur once as $(1, 2)$ differences. Similarly, we can verify that each element of M occurs once as a mixed difference. Hence the corollary is proved.

The results of the above corollaries are given by Bose (1939).

Van Lint (1973) gave various methods of constructing BIB designs with repeated sets. The following corollaries give some of his results.

Corollary 4.15.4 *A BIB design with repeated sets and with parameters*

$$v = 12t + 4, \quad b = (4t + 1)(12t + 4), \quad r = 12t + 3,$$
$$k = 3, \quad \lambda = 2, \quad b^* = (2t + 1)(12t + 4) \quad (4.52)$$

can be constructed from the $4t + 1$ initial sets

$$(0, 3t - i, 3t + 2 + i), \quad i = 0, 1, \ldots, t - 1 \text{ (twice each)}$$
$$(0, 5t + 1 - i, 5t + 2 + i), \quad i = 0, 1, \ldots, t - 1 \text{ (twice each)}$$
$$(0, 3t + 1, 6t + 2).$$

Corollary 4.15.5 *A BIB design with repeated sets and with parameters*

$$v = 6t + 2, \quad b = (6t + 1)(6t + 2), \quad r = 3(6t + 1),$$
$$k = 3, \quad \lambda = 6, \quad b^* = (t + 3)(6t + 1) \quad (4.53)$$

can be constructed, when $v - 1$ is a prime or a prime power, from the $6t + 2$ initial sets

(x^0, x^{2t}, x^{4t}), (five times)
$(x^i, x^{2t+i}, x^{4t+i})$, $i = 1, 2, \ldots, t - 1$ (six times each)
$(\infty, 0, x^s), (\infty, 0, x^{s+t}), (\infty, 0, x^{s+2t})$,

where x is a primitive root of $GF(v - 1)$ such that $x^{2t} - 1 = x^s$.

Corollary 4.15.6 *A BIB design with repeated sets and with parameters*

$$v = 12t + 6, \quad b = (4t + 2)(12t + 5), \quad r = 12t + 5,$$
$$k = 3, \quad \lambda = 2, \quad b^* = (2t + 2)(12t + 5), \tag{4.54}$$

can be constructed from the $4t + 2$ initial sets

$(0, 3t - i, 3t + 2 + i)$, $i = 0, 1, \ldots, t - 1$ (twice each)
$(0, 5t + 1 - i, 5t + 2 + i)$, $i = 0, 1, \ldots, t - 1$ (twice each)
$(0, 3t + 1, 6t + 2), (\infty, 0, 6t + 2)$.

Corollary 4.15.7 *A BIB design with repeated sets and with parameters*

$$v = 4t + 2, \quad b = (2t + 1)(4t + 1), \quad r = 8t + 2,$$
$$k = 4, \quad \lambda = 6, \quad b^* = (t + 2)(4t + 1), \tag{4.55}$$

can be constructed when $v - 1$ is a prime or a prime power and $t \not\equiv 2 \ (mod\ 3)$ from the $2t + 1$ initial sets

$(x^i, x^{t+i}, x^{2t+i}, x^{3t+i})$, $i = 1, 2, \ldots, t - 1$ (twice each),
$(x^0, x^t, x^{2t}, x^{3t}), (\infty, 0, x^t - 1, x^t + 1), (\infty, 0, x^t - 1, x^{2t})$,

where x is a primitive root of $GF(v - 1)$.

Let M be a module of v elements and let there exist one initial set $S = (a_1, a_2, \ldots, a_k)$ in which nonzero elements of M occur λ times as differences. That initial set is called a difference set, S. An integer $q \in M$ is called a multiplier of the difference set if

$$q(a_1, a_2, \ldots, a_k) = (a_1 q, a_2 q, \ldots, a_k q)$$
$$= (a_1 + q^*, a_2 + q^*, \ldots, a_k + q^*)$$
$$= (a_1, a_2, \ldots, a_k) + q^*,$$

where $q^* \in M$.

All known difference sets have nontrivial multipliers and for results on multipliers, the reader is referred to Baumert (1971), Hall (1967), Lander (1983), and Mann (1965).

4.7 A Statistical Model to Distinguish Designs with Different Supports

The standard statistical analysis will not distinguish between solutions of BIB designs with different support sizes. Raghavarao, Federer and Schwager (1986) discussed a statistical model that distinguished different solutions of a BIB design and that model has potential applications in market research and intercropping experiments. We will explain it in the context of market research.

Let A, B, C, D be four different brands of a product available in a store and Y_A be the revenue received for brand A in a given period. Then, we model

$$Y_A = \beta_A + \beta_{A(B)} + \beta_{A(C)} + \beta_{A(D)} + e_A, \tag{4.56}$$

where β_A is the effect of brand A, $\beta_{A(B)}$ is the competing effect (also called cross effect) of brand B on A, $\beta_{A(C)}$ is the competing effect of brand C on A, $\beta_{A(D)}$ is the competing effect of brand D on A, and e_A is random error distributed with mean 0 and variance σ^2. There are $v\,(v-2)$ independent competing effects elementary contrasts of the type $\beta_{i(j)} - \beta_{i(j')}$ for $i \neq j \neq j' \neq i$ while studying v brands. We will return to the optimal designs for this setting and further details in Chap. 6. If BIB designs with the model (4.56) are used, the number of independent estimable elementary contrasts for competing effects are different for different solutions with different support sizes.

There are 10 different solutions for a BIB design with parameters $v = 7, b = 21$, $r = 9, k = 3, \lambda = 3$ and Seiden (1977) showed that every solution is isomorphic to one of the 10 solutions. The solutions are listed in Table 4.3 with columns showing the support sizes, rows showing the sets and frequencies in the body of the table. The last row of the table gives the number of independent estimable contrasts of competing effects.

4.8 Non-Existence Results

The solution for a BIB design may not exist even if the parameters v, b, r, k and λ satisfy the necessary conditions:

$$vr = bk, \quad r(k-1) = \lambda(v-1), \quad b \geq v.$$

A powerful tool to show the non-existence of BIB designs and other configurations is the Hasse–Minkowski invariant using rational congruence of matrices. These results are well documented with detailed references in Raghavarao (1971, Chap. 12). We will only state the following theorem and illustrate it.

Table 4.3. Values of f_j[a] for all possible values of b^* for the BIB design with parameters $v = 7$, $b = 21, r = 9, k = 3$ and $\lambda = 3$.

Block Composition	Support Size (b^*)									
	7	11	13	14	15	17	18	19	20	21
1 2 3	3	3	3	2	3	2	2	3	2	1
1 2 4	–	–	–	1	–	1	–	–	–	1
1 2 5	–	–	–	–	–	–	–	–	–	1
1 2 6	–	–	–	–	–	–	–	–	1	–
1 2 7	–	–	–	–	–	–	1	–	–	–
1 3 4	–	–	–	–	–	–	–	–	1	–
1 3 5	–	–	–	–	–	–	–	–	–	–
1 3 6	–	–	–	1	–	1	1	–	–	1
1 3 7	–	–	–	–	–	–	–	–	–	1
1 4 5	3	3	–	2	1	2	2	1	1	1
1 4 6	–	–	1	–	1	–	–	1	–	–
1 4 7	–	–	2	–	1	–	1	1	1	1
1 5 6	–	–	2	–	1	–	1	1	1	1
1 5 7	–	–	1	1	1	1	–	1	1	–
1 6 7	3	3	–	2	1	2	1	1	1	1
2 3 4	–	–	–	–	–	–	–	–	–	1
2 3 5	–	–	–	–	–	1	1	–	–	–
2 3 6	–	–	–	–	–	–	–	–	–	–
2 3 7	–	–	–	1	–	–	–	–	1	1
2 4 5	–	–	1	–	1	–	–	1	1	–
2 4 6	3	2	2	2	2	1	2	1	1	1
2 4 7	–	1	–	–	–	1	1	1	1	–
2 5 6	–	1	–	1	–	1	1	1	1	1
2 5 7	3	2	2	2	2	1	1	1	1	1
2 6 7	–	–	1	–	1	1	–	1	–	1
3 4 5	–	–	2	1	1	–	1	1	1	1
3 4 6	–	1	–	–	–	1	1	1	1	1
3 4 7	3	2	1	2	2	2	1	1	–	–
3 5 6	3	2	1	2	2	1	–	1	1	1
3 5 7	–	1	–	–	–	1	1	1	1	1
3 6 7	–	–	2	–	1	–	1	1	1	–
4 5 6	–	–	–	–	–	1	–	–	–	–
4 5 7	–	–	–	–	–	–	–	–	–	1
4 6 7	–	–	–	1	–	–	–	–	1	1
5 6 7	–	–	–	–	–	–	1	–	–	–
#	14	20	25	28	26	31	32	28	34	35

[a] Value $f_j = 0$ is indicated by –.

\# Number of independent estimable elementary competing effects contrasts.

Theorem 4.16 *A symmetric BIB design with parameters $v = b, r = k, \lambda$ is non-existent if $(r - \lambda)$ is not a perfect square, when v is even. When v is odd, let g and h be the square free parts of $r - \lambda$ and λ. The design is non-existent if p is a prime such that*

a. *$p \mid g$, $p \nmid h$ and the equation $x^2 \equiv (-1)^{(v-1)/2}h \pmod{p}$ has no solution*
b. *$p \mid h$, $p \nmid g$ and the equation $x^2 \equiv g \pmod{p}$ has no solution*
c. *$p \mid g$ and $p \mid h$ and the equation $x^2 \equiv (-1)^{(v+1)/2}g_0 h_0 \pmod{p}$ has no solution, where $g = pg_0, h = ph_0$.*

The BIB design with parameters $v = 22 = b, r = 7 = k, \lambda = 2$ is non-existent, because $r - \lambda = 5$ is not a perfect square.

The BIB design with parameters $v = 43 = b, r = 7 = k, \lambda = 1$ is non-existent. Here $g = 6, h = 1$. Take $p = 3$ so that $p \mid g$. However, the equation

$$x^2 \equiv -1 \pmod{3}$$

has no solution.

4.9 Concluding Remarks

BIB designs are known in the literature from the 19th century. Steiner (1853) discussed the BIB designs with $k = 3$ and $\lambda = 1$ and they are known as Steiner triple systems. A bibliography and survey of such designs was given by Doyen and Rosa (1973, 1977).

Kirkman (1850) was interested in resolvable BIB designs through a school girls problem. Ray–Chaudhuri and Wilson (1971) completely solved the problem. S.S. Shrikhande (1976) and Baker (1983) gave a survey of results on affine resolvable BIB designs. Resolvable variance balanced designs are discussed by Mukerjee and Kageyama (1985). Also the review paper of M.S. Shrikhande (2001) discusses more results on α-resolvable and affine α-resolvable designs.

Yates (1936a) formally introduced BIB designs in agricultural experiments. Bose (1939) in a ground breaking paper discussed various construction methods of BIB designs, and he with his collaborators made significant contributions for the constructions and combinatorics of BIB designs.

The books by Beth, Jung-Nickel, Lenz (1999), Hall (1967), Raghavarao (1971), and Street and Street (1987) give excellent account of all aspects of these designs. We also suggest the volume edited by Colbourn and Dinitz (1996) as a valuable resource for the work on block designs and BIB designs in particular.

Table 4.4. Paramters of BIB designs with $v, b, \leq 100, r, k, \lambda \leq 15$ and their solutions.

Series	v	b	r	k	λ	Solution
1	3	3	2	2	1	Irreducible
2	4	6	3	2	1	Irreducible
3	4	4	3	3	2	Irreducible
4	5	10	4	2	1	Irreducible
5	5	5	4	4	3	Irreducible
6	5	10	6	3	3	Irreducible
7	6	15	5	2	1	Irreducible
8	6	10	5	3	2	Residual of Series 31
9	6	6	5	5	4	Irreducible
10	6	15	10	4	6	Irreducible
11	7	7	3	3	1	PG(2, 2): 1-flats as sets
12	7	7	4	4	2	Complement of Series 11
13	7	21	6	2	1	Irreducible
14	7	7	6	6	5	Irreducible
15	7	21	15	5	10	Irreducible
16	8	28	7	2	1	Irreducible
17	8	14	7	4	3	Difference set: (∞, x^0, x^2, x^4); $(0, x^1, x^3, x^5)$; $x \in \mathrm{GF}(7)$
18	8	8	7	7	6	Irreducible
19	9	12	4	3	1	EG(2, 3): 1-flats as sets
20	9	36	8	2	1	Irreducible
21	9	18	8	4	3	Difference set: (x^0, x^2, x^4, x^6); (x, x^1, x^3, x^5); $x \in \mathrm{GF}(3^2)$
22	9	12	8	6	5	Complement of Series 19
23	9	9	8	8	7	Irreducible
24	9	18	10	5	5	Complement of Series 21
25	10	15	6	4	2	Residual of Series 54
26	10	45	9	2	1	Irreducible
27	10	30	9	3	2	Difference set: $(0_1, 3_1, 1_2)$; $(1_1, 2_1, 1_2)$; $(1_2, 4_2, 4_1)$; $(0_2, 4_2, 1_2)$; $(0_1, 3_1, 4_2)$; $(1_1, 2_1, 4_2)$; mod 5
28	10	18	9	5	5	Residual of Series 63
29	10	15	9	6	5	Complement of Series 25
30	10	10	9	9	8	Irreducible
31	11	11	5	5	2	Difference set: $(x^0, x^2, x^4, x^6, x^8)$; $x \in \mathrm{GF}(11)$
32	11	11	6	6	3	Complement of Series 31
33	11	55	10	2	1	Irreducible
34	11	11	10	10	9	Irreducible
35	11	55	15	3	3	Difference set: $(0, x^0, x^5)$; $(0, x^1, x^6)$; $(0, x^2, x^7)$; $(0, x^3, x^8)$; $(0, x^4, x^9)$; $x \in \mathrm{GF}(11)$
36	12	44	11	3	2	Difference set: $(0, 1, 3)$; $(0, 1, 5)$; $(0, 4, 6)$; $(\infty, 0, 3)$; mod 11
37	12	33	11	4	3	Difference set: $(0, 1, 3, 7)$; $(0, 2, 7, 8)$; $(\infty, 0, 1, 3)$; mod 11
38	12	22	11	6	5	Difference set: $(0, 1, 3, 7, 8, 10)$; $(\infty, 0, 5, 6, 8, 10)$; mod 11
39	13	13	4	4	1	PG (2, 3): 1-flats as sets
40	13	26	6	3	1	Difference set: (x^0, x^4, x^8); (x, x^5, x^9); $x \in \mathrm{GF}(13)$

Table 4.4. (*Continued*)

Series	v	b	r	k	λ	Solution
41	13	13	9	9	6	Complement of Series 39
42	13	26	12	6	5	Difference set: $(x^0, x^2, x^4, x^6, x^8, x^{10})$; $(x^1, x^3, x^5, x^7, x^9, x^{11})$; $x \in GF(13)$
43	13	13	12	11	11	Irreducible
44	13	39	15	5	5	Difference set: $(0, x^0, x^3, x^6, x^9)$; $(0, x^1, x^4, x^7, x^{10})$; $(0, x^2, x^5, x^8, x^{11})$; $x \in GF(13)$
45	14	26	13	7	6	Residual of Series 81
46	14	14	13	13	12	Irreducible
47	15	35	7	3	1	Difference set: $(1_1, 4_1, 0_2)$; $(2_1, 3_1, 0_2)$; $(1_2, 4_2, 0_3)$; $(2_2, 3_2, 0_3)$;$(1_3, 4_3, 0_1)$;$(2_3, 3_3, 0_1)$;$(0_1, 0_2, 0_3)$; mod 5
48	15	15	7	7	3	Complement of Series 49
49	15	15	8	8	4	Theorem 4.13 on Series 7
50	15	35	14	6	5	Difference set: $(\infty, 0_1, 0_2, 1_2, 2_2, 4_2)$; $(\infty, 0_1, 3_1, 5_1, 6_1, 0_2)$; $(0_1, 1_1, 3_1, 0_2, 2_2, 6_2)$; $(0_1, 1_1, 3_1, 1_2, 5_2, 6_2)$; $(0_1, 4_1, 5_1, 0_2, 1_2, 3_2)$; mod 7
51	15	42	14	5	4	$(0, 1, 4, 9, 11)$; $(0, 1, 4, 10, 12)$; $(\infty, 0, 1, 2, 7)$; mod 14
52	15	15	14	14	13	Irreducible
53	16	20	5	4	1	EG $(2, 4)$: 1-flats as sets
54	16	16	6	6	2	Complement of Series 56
55	16	24	9	6	3	Residual of Series 77
56	16	16	10	10	6	Theorem 4.12 on Series 3 with itself
57	16	80	15	3	2	Difference set: (x^0, x^5, x^{10}); (x^1, x^6, x^{11}); (x^2, x^7, x^{12}); (x^3, x^8, x^{13}); (x^2, x^7, x^{12}); (x^3, x^8, x^{13}); (x^4, x^9, x^{14}); $x \in GF(2^4)$
58	16	48	15	5	4	Difference set: $(x^0, x^3, x^6, x^9, x^{12})$; $(x^1, x^4, x^7, x^{10}, x^{13})$; $(x^2, x^5, x^8, x^{11}, x^{14})$; $x \in GF(2^4)$
59	16	40	15	6	5	Sets of Series 54 and 55 together
60	16	30	15	8	7	EG$(4, 2)$; 3-flats as sets
61	16	16	15	15	14	Irreducible
62	19	57	9	3	1	Difference set: (x^0, x^6, x^{12}); (x^1, x^7, x^{13}); (x^2, x^8, x^{14}); $x \in GF(19)$
63	19	19	9	9	4	Difference set: $(x^0, x^2, x^4, x^6, x^8, x^{10}, x^{12}, x^{14}, x^{16})$; $x \in GF(19)$
64	19	19	10	10	5	Complement of Series 63
65	19	57	12	4	2	Difference set: $(0, x^0, x^6, x^{12})$; $(0, x^1, x^7, x^{13})$; $(0, x^2, x^8, x^{14})$; $x \in GF(19)$
66	21	21	5	5	1	PG$(2, 4)$: 1-flats as sets
67	21	70	10	3	1	Difference set: $(1_1, 6_1, 0_2)$; $(2_1, 5_1, 0_2)$; $(3_1, 4_1, 0_2)$; $(1_2, 6_2, 0_3)$; $(2_2, 5_2, 0_3)$; $(3_2, 4_2, 0_3)$; $(1_3, 6_3, 0_1,)$; $(2_3, 5_3, 0_1)$; $(3_3, 4_3, 0_1)$; $(0_1, 0_2, 0_3)$; mod 7

Table 4.4. (*Continued*)

Series	v	b	r	k	λ	Solution
68	21	30	10	7	3	Residual of Series 87
69	21	42	12	6	3	Difference set: $(0_1, 5_1, 1_2, 4_2, 2_3, 3_3)$; $(0_1, 1_1, 3_1, 0_2, 1_2, 3_2)$; $(0_2, 5_2, 1_3, 4_3, 2_1, 3_1)$; $(0_2, 1_2, 3_2, 0_3, 1_3, 3_3)$; $(0_3, 5_3, 1_1, 4_1, 2_2, 3_2)$; $(0_3, 1_3, 3_3, 0_1, 1_1, 3_1)$; mod 7
70	21	35	15	9	6	Residual of Series 92
71	22	44	14	7	4	$(0, 6, 11, 15, 18, 20, 21)$; $(0, 5, 7, 8, 9, 13, 19)$; mod 22
72	22	77	14	4	2	Difference set: $(x_1^0, x_1^3, x_2^a, x_2^{a+3})$; $(x_1^1, x_1^4, x_2^{a+1}, x_2^{a+4})$; $(x_1^2, x_1^5, x_2^{a+2}, x_2^{a+5})$; $(x_2^0, x_2^3, x_3^a, x_3^{a+3})$; $(x_2^1, x_2^4, x_3^{a+1}, x_3^{a+4})$; $(x_2^2, x_2^5, x_3^{a+2}, x_3^{a+5})$; $(x_3^0, x_3^3, x_1^a, x_1^{a+3})$; $(x_3^1, x_3^4, x_1^{a+1}, x_1^{a+4})$; $(x_3^2, x_3^5, x_1^{a+2}, x_1^{a+5})$; $(\infty, 0_1, 0_2, 0_3)$; $(\infty, 0_1, 0_2, 0_3)$; $x \in GF(7)$
73	23	23	11	11	5	Difference set: $(x^0, x^2, x^4, x^6, x^8, x^{10}, x^{12}, x^{14}, x^{16}, x^{18}, x^{20})$; $x \in GF(23)$
74	23	23	12	12	6	Complement of 73
75	25	30	6	5	1	EG(2, 5): 1-flats as sets
76	25	50	8	4	1	Difference set: $(0, x^0, x^8, x^{16})$, $(0, x^2, x^{10}, x^{18})$; $x \in GF(5^2)$
77	25	25	9	9	3	Trial-and-error solution; refer to Fisher and Yates (1963) statistical tables
78	25	100	12	3	1	Difference set: (x^0, x^8, x^{16}); (x^1, x^9, x^{17}), (x^2, x^{10}, x^{18}); (x^3, x^{11}, x^{19}); $x \in GF(5^2)$
79	26	65	15	6	3	Difference set: $(0_1, 3_1, 1_2, 2_2, 0_3, 3_3)$; $(0_1, 3_1, 1_3, 2_3, 1_4, 2_4)$; $(0_2, 3_2, 1_3, 2_3, 0_4, 3_4)$; $(0_2, 3_2, 1_4, 2_4, 1_5, 2_5)$; $(0_3, 3_3, 1_4, 2_4, 0_5, 3_5)$; $(0_3, 3_3, 1_5, 2_5, 1_1, 2_1)$; $(0_4, 3_4, 1_5, 2_5, 0_1, 3_1)$; $(0_4, 3_4, 1_1, 2_1, 1_2, 2_2)$; $(0_5, 3_5, 1_1, 2_1, 0_2, 3_2)$; $(0_5, 3_5, 1_2, 2_2, 1_3, 2_3)$; $(\infty, 0_1, 3_2, 1_3, 4_4, 2_5)$; $(\infty, 0_1, 2_2, 4_3, 1_4, 3_5)$; $(\infty, 0_1, 0_2, 0_3, 0_4, 0_5)$; mod 5
80	27	39	13	9	4	EG(3, 3): 2-flats as sets
81	27	27	13	13	6	Difference set: $(x^0, x^2, x^4, x^6, x^8, x^{10}, x^{12}, x^{14}, x^{16}, x^{18}, x^{20}, x^{22}, x^{24})$; $x \in GF(3^3)$
82	27	27	14	14	7	Complement of 81
83	28	63	9	4	1	Difference set: $(x_1^0, x_1^4, x_2^a, x_2^{a+4})$; $(x_1^2, x_1^6, x_2^{a+2}, x_2^{a+6})$; $(x_2^0, x_2^4, x_3^a, x_3^{a+4})$; $(x_2^2, x_2^6, x_3^{a+2}, x_3^{a+6})$; $(x_3^0, x_3^4, x_1^a, x_1^{a+4})$; $(x_3^2, x_3^6, x_1^{a+2}, x_1^{a+6})$; $(\infty, 0_1, 0_2, 0_3)$; $x \in GF(3^2)$

Table 4.4. (*Continued*)

Series	v	b	r	k	λ	Solution
84	28	36	9	7	2	Residual of Series 93
85	29	58	14	7	3	Difference set: $(x^0, x^4, x^8, x^{12}, x^{16}, x^{20}, x^{24})$: $(x^1, x^5, x^9, x^{13}, x^{17}, x^{21}, x^{25})$; $x \in \mathrm{GF}(29)$
86	31	31	6	6	1	PG(2, 5): 1-flats as sets
87*	31	31	10	10	3	Difference set: $(A, 0_1, 5_1, 1_2, 4_2, 2_3, 3_3, 2_4, 4_4, 5_4)$; $(B, 0_1, 3_1, 1_2, 2_2, 4_3, 6_3, 1_4, 3_4, 4_4)$; $(C, 0_1, 1_1, 3_2, 5_2, 2_3, 6_3, 0_4, 2_4, 3_4)$; $(0_1, 1_1, 3_1, 0_2, 1_2, 3_2, 0_3, 1_3, 3_3, 6_4)$; mod 7; $(A, B, C, 0_1, 1_1, 2_1, 3_1, 4_1, 5_1, 6_1)$; $(A, B, C, 0_2, 1_2, 2_2, 3_2, 4_2, 5_2; 6_2)$; $(A, B, C, 0_3, 1_3, 2_3, 3_3, 4_3, 5_3, 6_3)$;
88	31	93	15	5	2	Difference set: $(x^0, x^6, x^{12}, x^{18}, x^{24})$: $(x^1, x^7, x^{13}, x^{19}, x^{25}), (x^2, x^8, x^{14}, x^{20}, x^{26})$; $x \in \mathrm{GF}(31)$
89	31	31	15	15	7	PG(4, 2): 3-flats as sets
90	33	44	12	9	3	Residual of Series 99
91	36	84	14	6	2	$(0, 1, 3, 5, 11, 23); (0, 5, 8, 9, 18, 24);$ $(\infty, i, i+7, i+14, i+21, i+28), i = 0, 1, \ldots, 6$ (each of these 7 sets taken twice); mod 35
92	36	36	15	15	6	Wallis (1969)
93	37	37	9	9	2	Difference set: $(x^0, x^4, x^8, x^{12}, x^{16}, x^{20}, x^{24}, x^{28}, x^{32})$; $x \in \mathrm{GF}(37)$
94	40	40	13	13	4	PG(3, 3): 2-flats as sets
95	41	82	10	5	1	Difference set: $(x^0, x^8, x^{16}, x^{24}, x^{32})$: $(x^2, x^{10}, x^{18}, x^{26}, x^{34})$; $x \in \mathrm{GF}(41)$
96	43	86	14	7	2	Difference set: $(0_1, 1_2, 6_2, 5_3, 2_3, 3_4, 4_4)$; $(0_1, 1_3, 6_3, 5_4, 2_4, 3_5, 4_5); (0_1, 1_4, 6_4, 5_5, 2_5, 3_6, 4_6)$; $(0_1, 1_5, 6_5, 5_6, 2_6, 3_2, 4_2); (0_1, 1_6, 6_6, 5_2, 2_2, 3_3, 4_3)$; $(0_1, 1_2, 6_2, 5_5, 2_5, 3_3, 4_3); (0_1, 1_3, 6_3, 5_6, 2_6, 3_4, 4_4)$; $(0_1, 1_4, 6_4, 5_2, 2_2, 3_5, 4_5); (0_1, 1_5, 6_5, 5_3, 2_3, 3_6, 4_6)$; $(0_1, 1_6, 6_6, 5_4, 2_4, 3_2, 4_2); (0_1, 0_2, 0_3, 0_4, 0_5, 0_6, \infty)$ (twice); mod 7; $(0_1, 1_1, 2_1, 3_1, 4_1, 5_1, 6_1)$ (twice)
97	45	55	11	9	2	Residual of Series 101
98	45	99	11	5	1	Difference set: $(x_1^0, x_1^4, x_3^a, x_3^{a+4}, 0_2)$; $(x_1^2, x_1^6, x_3^{a+2}, x_3^{a+6}; 0_2); (x_2^0, x_2^4, x_4^a, x_4^{a+4}, 0_3)$; $(x_2^2, x_2^6, x_4^{a+2}, x_4^{a+6}; 0_3); (x_3^0, x_3^4, x_5^a, x_5^{a+4}; 0_4)$; $(x_3^2, x_3^6, x_5^{a+2}, x_5^{a+6}; 0_4); (x_4^0, x_4^4, x_1^a, x_1^{a+4}, 0_5)$; $(x_4^2, x_4^6, x_1^{a+2}, x_1^{a+6}; 0_5); (x_5^0, x_5^4, x_2^a, x_2^{a+4}, 0_1)$; $(x_5^2, x_5^6, x_2^{a+2}, x_2^{a+6}; 0_1); (0_1, 0_2, 0_3, 0_4, 0_5); x \in \mathrm{GF}(3^2)$
99	45	45	12	12	3	Wallis (1969)

Table 4.4. (*Continued*)

Series	v	b	r	k	λ	Solution
100	49	56	8	7	1	EG(2, 7): 1-flats as sets
101	56	56	11	11	2	Hall, Jr., Lane and Wales (1970)
102	57	57	8	8	1	PG(2, 7): 1-flats as sets
103	64	72	9	8	1	EG(2, 8): 1-flats as sets
104	66	78	13	11	2	Residual of Series 107
105	71	71	15	15	3	Becker and Haemaers (1980)
106	73	73	9	9	1	PG(2, 8): 1-flats as sets
107	79	79	13	13	2	Aschbacher (1971)
108	81	90	10	9	1	EG(2, 9): 1-flats as sets
109	91	91	10	10	1	PG(2, 9): 1-flats as sets

*The first four sets should be developed mod 7, while keeping A, B, C unaltered in the development. Thus we get 28 sets, and the last three complete the solution.

$x \in$ GF(s) should be interpreted as x is a primitive root of GF(s), and $a \neq 0$.

Mohan, Kageyama and Nair (2004) classified symmetric BIB designs into three types: Type I with $k = n\lambda$ for $n \geq 2$, Type II with $k = n\lambda + 1$ for $n \geq 1$, and Type III with $k = n\lambda + m$ for $n \geq 1$, $m \geq 2$. They characterized the parameters for these types and gave a list of parameters in the range $v = b \leq 111$, $r \leq 55$, $\lambda \leq 30$ indicating the (non)existence status of corresponding designs.

The association matrices that we introduce in Chap. 8 are useful to construct BIB designs. Such procedures were studied by Blackwelder (1969), Kageyama (1974a), Mukhopadhyay (1974), and Shrikhande and Singh (1962).

Attempts are being made to show the sufficiency for the conditions on the parameters of BIB designs and Wilson (1972a, b, 1974) established the following remarkable theorem:

Theorem 4.17 *For a given k, there exists a constant $c(k)$ such that the BIB design with parameters v, b, r, k, λ satisfying $vr = bk$, $\lambda(v-1) = r(k-1)$, $b \geq v$ exists, whenever $v > c(k)$.*

Hanani (1975) in a lengthy article reviewed and updated the recursive constructions of these designs for small k. Except the design with parameters $v = 15$, $b = 21$, $r = 7$, $k = 5$, $\lambda = 2$, all other designs exist for $k = 5$.

The symmetric BIB designs with parameters $v = 4u^2$, $k = 2u^2 + u$, $\lambda = u^2 + u$ were constructed for many values of u (see Koukouvinas, Koumias, Sebery, 1989; Sebery, 1991; and Xia, Xia, and Sebery, 2003).

Some recursive construction methods of affine resolvable designs are given by Agrawal and Boob (1976), and Griffiths and Mavron (1972).

Table 4.5. Parametric combinations in the range $v, b \leq 100, r, k \leq 15$ with unknown solutions.

Series	v	b	r	k	λ
1	22	33	12	8	4
2	28	42	15	10	5
3	40	52	13	10	3
4	46	69	9	6	1
5	46	69	15	10	3
6	51	85	10	6	1
7	56	70	15	12	3

We give an updated list of parametric combinations of BIB designs in the range $v, b \leq 100, r, k \leq 15$ in Table 4.4. The method of construction or the source is listed against each parametric combination. Designs obtainable by duplicating known designs are not included in the list. The list of parametric combinations in the range $v, b \leq 100, r, k \leq 15$ whose solutions yet seem to be unknown to the authors are given in Table 4.5.

5

Balanced Incomplete Block Designs — Applications

5.1 Finite Sample Support and Controlled Sampling

Let us consider a finite population of N units and a sample of size n is selected from that population without replacement by some probability mechanism. Hortvitz–Thompson estimator (1952) of the population total is based on the first-order inclusion probability, π_i, and Yates–Grundy estimator (1953) of its variance is based upon π_i, and the second-order inclusion probability, π_{ij}. Here π_i is the probability that the ith population unit is included in the sample, and π_{ij} is the joint probability that the ith and jth units are both included in the sample.

In simple random sample without replacement, the support of the sample consists of all possible combinations of n units taken from N units and the support size is $\binom{N}{n}$. If we write the samples as sets and units as symbols, the support is the irreducible BIB design with parameters

$$v = N, \quad b = \binom{N}{n}, \quad r = \binom{N-1}{n-1}, \quad k = n, \quad \lambda = \binom{N-2}{n-2}. \tag{5.1}$$

Now $\pi_i = n/N$, and $\pi_{ij} = \{n(n-1)\}/\{N(N-1)\}$ for $i \neq j; i, j = 1, 2, \ldots, N$.

Chakrabarti (1963b) noted that the support of the simple random sample without replacement can be considerably reduced by taking any BIB design with $b < \binom{N}{n}$, if it exists and the blocks are distinct. His results were extended to designs with repeated blocks by Foody and Hedayat (1977), and Wynn (1977).

Let a BIB design exist with $v = N, k = n$, possibly with repeated blocks with support size b^*, which is much smaller than $\binom{N}{n}$. Let $\mathbf{f}' = (f_1, f_2, \ldots, f_{b^*})$ be the frequency vector and the number of sets of the BIB design is $b = \sum_{i=1}^{b^*} f_i$. Using a probability mechanism, select the ith distinct set of the design as a sample with probability f_i/b. The units corresponding to the symbols of the selected set is our sample. This sample gives inclusion probabilities

$$\pi_i = r/b = rv/bv = k/v = n/N, \tag{5.2}$$

$$\pi_{ij} = \lambda/b = \{\lambda(v-1)k\}/\{b(v-1)k\}$$
$$= \{k(k-1)\}/\{v(v-1)\} = \{n(n-1)\}/\{N(N-1)\},$$

which are the same as the inclusion probabilities of a simple random sample without replacement. Thus this scheme is equivalent to the commonly used simple random sample without replacement.

In practice, all the $\binom{N}{n}$ possible samples will not be equally desirable. When the units of the sample are spread out demographically, it is more expensive to get information from all units in the selected sample. In this sense, some samples are preferred samples, while others are nonpreferred samples. We can assign very low probability for nonpreferred samples and give high probability for preferred samples to be selected by using the support of the sets of a BIB design with $v = N$, $k = n$. The inclusion probabilities can be easily calculated and the estimators obtained easily. This is a controlled sampling procedure discussed by Avadhani and Sukhatme (1973). Rao and Nigam (1990) used linear programming formulation to controlled sampling. They determined the probabilities of selecting the sets of the design with parameters (5.1), so that the second-order inclusion probabilities are equal to the simple random sample without replacement, while minimizing the total probability of selecting nonpreferred samples. See also, Rao and Vijayan (2001).

5.2 Randomized Response Procedure

In surveys, the respondents hesitate and do not respond to questions which are sensitive as people do not divulge personal secrets before strangers. To overcome this problem and estimate the proportion of people with a sensitive characteristic, Warner (1965) developed an ingenious procedure. He suggested making two statements:

1. I belong to the sensitive category
2. I do not belong to the sensitive category

and the respondent chooses statement 1 or 2 with a preselected probability p or $1 - p$ and simply answers yes or no without telling the investigator whether he/she is answering statement 1 or 2. Since the interviewer does not know whether the answer is for statement 1 or 2, the anonymity of the response is somewhat protected and to protect it to the maximum extent, p is selected close to $1/2$, but $p \neq 1/2$. Based on the yes answers, the proportion of people belonging to the sensitive category can be estimated. Several alternatives and improvements were suggested, and the monograph by Chaudhuri and Mukerjee (1988) discusses the developments.

In this section, we will give the application of BIB designs to estimate the proportions in one or several sensitive categories, a method developed by Raghavarao and Federer (1979). Suppose that there are m sensitive categories and we want to estimate the population proportions π_i for each of the m sensitive categories. To this set of m sensitive questions, we add $v - m$ unrelated non-sensitive binary response questions. Of the v total binary response questions, we take subsets of k questions, where each set has not all sensitive questions. The respondents will be divided into groups with equal number of respondents in each group, and the number of groups equal to the number of subsets of k questions. Each respondent in the ith group will give a total response of all questions included in the subset without divulging the answer to each question in the subset and coding the responses no (yes) by 0 (1). This protects the anonymity to a larger extent, but not fully. The response total of 0 or k identifies all the responses of such respondents; otherwise, the individual responses precisely are not known. Let $\bar{Y}_i$ be the mean of all responses of all respondents answering the ith subset of questions, S_i. Then

$$E(\bar{Y}_i) = \sum_{\ell \in S_i} \pi_\ell, \quad i = 1, 2, \dots, b, \tag{5.3}$$

where $E(\bullet)$ is the expected value of the random variable in parentheses, π_ℓ is the proportion of people belonging to the category of ℓth question. The subsets have to be made such that all v proportions of m sensitive and $(v - m)$ non-sensitive questions are estimated.

This problem is the well-known spring balance weighing design problem without bias discussed by Mood (1946), and Raghavarao (1971). The optimal design is the Hadamard design, which is a BIB design with parameters

$$v = 4t - 1 = b, \quad r = 2t = k, \quad \lambda = t. \tag{5.4}$$

Thus, we select $4t - 1$ total sensitive and non-sensitive questions and form $4t - 1$ subsets each with $2t$ questions, where the subsets are made based on the solution of the BIB design with parameters (5.4). The n respondents will be divided into $4t - 1$ groups of u each ($n = (4t - 1)u$) and each of the u respondents will give a sum total response for the $2t$ questions of the ith set for $i = 1, 2, \dots, 4t - 1$. $\mathrm{Var}(\bar{Y}_i)$ for $i = 1, 2, \dots, 4t - 1$ are not all the same. However, since there are $4t - 1$ observations and $4t - 1$ parameters, the weighted least squares estimate π_ℓ is the same as solving the equations equating $\bar{Y}_i$ to its expected value given by

(5.3). $\mathrm{Var}\left(\bar{Y}_i\right)$ can be estimated from the individual u responses to S_i and this can be used to estimate the variance of the estimated π_ℓ.

This method was used by Smith, Federer and Raghavarao (1974) to estimate the proportion of students who cheated on tests, stole money, and took drugs. We will illustrate the method with a small artifical example.

Suppose we are interested to estimate the proportion of workers in a large company satisfied with the working conditions at the company. We consider 3 items:

> A. I am satisfied with the working conditions
>
> > No (0),   Yes (1)
>
> B. I eat my lunch in the company cafeteria
>
> > No (0),   Yes (1)
>
> C. I exercise at least once a week
>
> > No (0),   Yes (1).

We form 3 sets of a BIB design with parameters $v = 3 = b, r = 2 = k, \lambda = 1$. We randomly choose 15 employees, divide them randomly into 3 groups of 5 each, and request each employee to give a total score for the 2 items in his/her questionnaire. The following are the artificial data and summary statistics.

Set	Responses	Means $\left(\bar{Y}_i\right)$	Variances $\left(s_i^2\right)$
A, B	0, 1, 1, 2, 2	1.2	0.7
A, C	0, 0, 1, 1, 1	0.6	0.2
B, C	1, 1, 2, 2, 2	1.6	0.3

If π_i is the proportion of yes answers for ith item, $i = A, B, C$, we have

$$\hat{\pi}_A + \hat{\pi}_B = \bar{Y}_1 = 1.2; \quad \hat{\pi}_A + \hat{\pi}_C = \bar{Y}_2 = 0.6; \quad \hat{\pi}_B + \hat{\pi}_C = \bar{Y}_3 = 1.6.$$

Solving these equations and finding their estimated variances, we get

$$\hat{\pi}_A = \frac{\bar{Y}_1 + \bar{Y}_2 - \bar{Y}_3}{2} = 0.1; \quad \hat{\mathrm{Var}}(\hat{\pi}_A) = \frac{1}{4} \frac{\left(s_1^2 + s_2^2 + s_3^2\right)}{5} = 0.6,$$

$$\hat{\pi}_B = \frac{\bar{Y}_1 - \bar{Y}_2 + \bar{Y}_3}{2} = 1.1; \quad \hat{\mathrm{Var}}(\hat{\pi}_B) = \frac{1}{4} \frac{\left(s_1^2 + s_2^2 + s_3^2\right)}{5} = 0.6.$$

Since $\hat{\pi}_B > 1$, we take $\hat{\pi}_B = 1$.

$$\hat{\pi}_C = \frac{-\bar{Y}_1 + \bar{Y}_2 + \bar{Y}_3}{2} = 0.5; \quad \hat{\mathrm{Var}}(\hat{\pi}_C) = \frac{1}{4}\frac{\left(s_1^2 + s_2^2 + s_3^2\right)}{5} = 0.6.$$

From this data, we infer that 10% of the workers are satisfied with the working conditions in the company.

5.3 Balanced Incomplete Cross Validation

Cross-validation is an important tool in model selection based on the predictive ability of the model. The available data consisting of n observations, will be split into two sets of n_{co} and n_{va} observations ($n_{co} + n_{va} = n$). The model will be constructed from n_{co} observations and will be validated on the n_{va} observations. If $\mathbf{Y}^{(2)}$ is a vector of n_{va} responses corresponding to the validation data, and if $\hat{\mathbf{Y}}^{(2)}$ is the estimated vector of responses for the validation data based on the model constructed from n_{co} observations, the average squared prediction error with this splitting is

$$\frac{\left(\mathbf{Y}^{(2)} - \hat{\mathbf{Y}}^{(2)}\right)'\left(\mathbf{Y}^{(2)} - \hat{\mathbf{Y}}^{(2)}\right)}{n_{va}}. \tag{5.5}$$

The average squared prediction error will be calculated for all $\binom{n}{n_{va}}$ combinations or fewer combinations. The selected model will have smallest sum of averaged squared prediction errors.

Shao (1993) suggested the use of BIB designs to consider the splitting of the data into construction and validation sets and he called it the Balanced Incomplete Cross-Validation Method.

Suppose there exists a BIB design with parameters $v = n$, $k = n_{va}$, r, b, λ. Each set of the BIB design provides a splitting of the sample data into construction and validation sets. The symbols in the set of the BIB design provide the data units for validation, whereas the symbols in the complementary set provide the units to construct the model.

Shao (1993) showed that the model selected by this method is asymptotically correct if $n_{co} \to \infty$ and $n_{va}/n \to 1$.

5.4 Group Testing

If we want to test a blood sample (or item) for the presence of a rare trait (or characteristic), the test result can be negative (positive) indicating the absence (presence) of the trait. We can also combine different samples and perform the test.

If the test result is negative, the trait is absent from each sample included in the test. If the test result is positive, at least one sample included in the test has the trait and it is not known which ones possess the trait. Further group tests, or individual tests are needed to identify the samples for the presence of the trait. The tests can be done sequentially using the information obtained earlier to plan the subsequent experiments and this strategy is called adaptive testing. A non-adaptive procedure plans all the tests simultaneously and identifies the samples (items) where the trait is present. The number of samples with the trait may or may not be known to the experimenter. In binomial testing, the number is unknown while in hypergeometric testing, the number is known. This area of research is called group testing. By group testing one can cut down the cost and time of screening to identify the samples with specific traits of interest.

Dorfman (1943) is the first to introduce this concept of group testing. During the Second World War, the army recruits were tested for syphilis. The blood samples were pooled and tested for syphilis and if the test is negative, all the recruits whose blood samples are included in the test were considered syphilis free. When the test result is positive, each blood sample is tested to determine who had syphilis. Late Professor Milton Sobel and his co-workers made pioneering work in adaptive methods. The non-adaptive testing is interesting from block designs perspective. An excellent treatment of this topic can be found in the monograph by Du and Hwang (1993), and the paper by Macula and Rykov (2001).

In this monograph we will be considering non-adaptive hypergeometric group testing problem.

Bush *et al.* (1984) introduced d-completeness in block designs according to Definition 5.1.

Definition 5.1 A block design in v symbols and b blocks is said to be *d-complete*, if given any d symbols $(\theta_1, \theta_2, \ldots, \theta_d)$, the union of the sets of the block design where none of the θ_i for $i = 1, 2, \ldots, d$ appear, is the set of v symbols excluding $\theta_1, \theta_2, \ldots, \theta_d$ of cardinality $v - d$.

If it is known that d of the v samples (items) have the trait, they can be tested and identified in b tests corresponding to a d-complete design. The samples included in the union of tests giving negative result do not have the trait, whereas the others possess the trait under investigation. In order for the group testing method to-be useful, v must be large and b must be very small in a d-complete design.

Bush *et al.* (1984) proved.

Theorem 5.1 *A BIB design with parameters v, b, r, k and λ is d-complete, if*

$$r - d\lambda > 0. \tag{5.6}$$

Proof. Given any d symbols $\theta_1, \theta_2, \ldots, \theta_d$, a symbol $\theta \neq \theta_i, i = 1, 2, \ldots, d$ can occur with θ_i in at most $d\lambda$ sets and when $r - d\lambda > 0$, θ occurs in a set without any of $\theta_1, \theta_2, \ldots, \theta_d$. Hence θ belongs to the union of the sets of the design where none of θ_i for $i = 1, 2, \ldots, d$ appear, if $r > d\lambda$.

The d-complete BIB designs as such are not of much use in group testing v samples in b tests because $b \geq v$. However, if n samples are to be tested and if $n = vw$, for large w, then the n samples can be divided into v groups of w samples and the ith group can be considered as the ith symbol of the d-complete BIB design and tests can be performed according to the sets of the BIB design. If we assume that each of the d groups of the samples have at least one sample with the abnormal trait, whereas the other $v - d$ groups of the samples are free of the abnormal trait, the d groups can be identified by group testing with the d-complete BIB design, and the dw individual samples can be tested to identify the defective items.

The concept of dual design is also helpful in this context. We will introduce the idea here and will pursue its detailed study in Chap. 7.

Definition 5.2 Given a block design, D, with v symbols in b sets, the *dual design, D^**, is obtained by interchanging the roles of sets and symbols in D. D^* will have b symbols arranged in v sets.

For example, u_1, u_2, u_3, u_4, are 4 symbols arranged in 6 sets $S_1, S_2, \ldots, S_6$ as follows in D:

$$S_1 : (u_1, u_2); \quad S_2 : (u_1, u_3); \quad S_3 : (u_1, u_4); \quad S_4 : (u_2, u_3);$$
$$S_5 : (u_2, u_4); \quad S_6 : (u_3, u_4).$$

In the dual design D^*, u_1, will be a set consisting of the symbols S_1, S_2, S_3 as u_1 occurred in the sets S_1, S_2, S_3 of D. Following this manner, we get the dual design D^* in 6 symbols $S_1, S_2, \ldots, S_6$ arranged in 4 sets u_1, u_2, u_3, and u_4 as follows:

$$u_1 : (S_1, S_2, S_3); \quad u_2 : (S_1, S_4, S_5); \quad u_3 : (S_2, S_4, S_6); \quad u_4 : (S_3, S_5, S_6).$$

The following theorem can be easily verified.

Theorem 5.2 *In a BIB design without repeated sets, if no set is a subset of the union of any other two sets, then its dual design is 2-complete.*

Theorem 5.2 provides many useful group testing designs to identify 2 samples with the abnormal trait. The duals of Steiner triple systems are useful group testing designs to identify 2 samples with the abnormal traits.

Weideman and Raghavarao (1987a, b) studied the group testing designs in detail under some assumptions and their constructions are based on Partially Balanced

Incomplete Block (PBIB) designs and BIB designs and we will return to their results in Chap. 8.

Schultz, Parnes and Srinivasan (1993) noted that d-complete designs may alert the experimenter if more than d samples with abnormal trait exist among the v tested samples. This is the case, when the union of the sets with negative test results contain less than $v - d$ symbols.

5.5 Fractional Plans to Estimate Main Effects and Two-Factor Interactions Inclusive of a Specific Factor

Let us consider a 2^n factorial experiment with n factors each at 2 levels. Sometimes the experimenter may be interested in estimating all the n main effects and $n - 1$, two-factor interactions involving a specific factor with each of the other $n - 1$ factors. Resolution IV fractional plans give the necessary designs; however, Damaraju and Raghavarao (2002) gave an interesting application of BIB designs to this problem.

Let D be a BIB design with parameters $v = n, b, r, k, \lambda$ and $\bar{D}$ be its complement design with parameters $\bar{v} = n, \bar{b} = b, \bar{r} = b - r, \bar{k} = n - k, \bar{\lambda} = b - 2r + \lambda$. It was shown in Chap. 4 that every triplet of distinct symbols occurs in $b - 3r + 3\lambda$ sets of D and $\bar{D}$. Let N be the incidence matrix of D and $\mathbf{n}_1, \mathbf{n}_2, \ldots, \mathbf{n}_v$ be the v columns of N'. Put $\mathbf{x}_i = 2\mathbf{n}_i - \mathbf{1}_b, \bar{\mathbf{x}}_i = \mathbf{1}_b - 2\mathbf{n}_i$.

Let $a_1, a_2, \ldots, a_n$ be the n factors and we are interested in estimating the main effects $A_1, A_2, \ldots, A_n$, and without loss of generality the two-factor interactions $A_1 A_2, A_1 A_3, \ldots, A_1 A_n$. We consider the $2b$ sets of D and $\bar{D}$ as $2b$ runs of the factorial experiment interpreting the presence (absence) of a symbol in the set as the high (low) level of the corresponding factor in the run.

The $2b \times 2n$ design matrix is then

$$X = \begin{pmatrix} \mathbf{1}_b\, \mathbf{x}_1\, \mathbf{x}_2 \ldots \mathbf{x}_n\, \mathbf{x}_1 \bullet \mathbf{x}_2\, \mathbf{x}_1 \bullet \mathbf{x}_3 \ldots \mathbf{x}_1 \bullet \mathbf{x}_n \\ \mathbf{1}_b\, \bar{\mathbf{x}}_1\, \bar{\mathbf{x}}_2 \ldots \bar{\mathbf{x}}_n\, \bar{\mathbf{x}}_1 \bullet \bar{\mathbf{x}}_2\, \bar{\mathbf{x}}_1 \bullet \bar{\mathbf{x}}_3 \ldots \bar{\mathbf{x}}_1 \bullet \bar{\mathbf{x}}_n \end{pmatrix} \tag{5.7}$$

and the observational setup can be written as

$$\mathbf{Y} = X\boldsymbol{\beta} + \mathbf{e}, \tag{5.8}$$

where $\mathbf{Y}$ is the response vector for the $2b$ runs,

$$\boldsymbol{\beta}' = (\mu, A_1, A_2, \ldots, A_n, A_1 A_2, A_1 A_3, \ldots, A_1 A_n),$$

μ is the general mean, $\bullet$ is the Hadamard product of vectors showing the term by term multiplication, and $\mathbf{e}$ is the vector of random errors distributed independently $N(0, \sigma^2)$.

Here

$$X'X = \begin{pmatrix} 2b & 0 & 2(b-4r+4\lambda)1'_{r-1} \\ 0 & 8(r-\lambda)I_n + 2(b-4r+4\lambda)J_n & 0 \\ 2(b-4r+4\lambda)1_{r-1} & 0 & 8(r-\lambda)I_{n-1} + 2(b-4r+4\lambda)J_{n-1} \end{pmatrix},$$

where 0 is a vector or matrix of appropriate order.

$X'X$ can easily be shown to be non-singular, and the main effects and the two-factor interactions can easily be estimated by the methods discussed in Chap. 1. Damaraju and Raghavarao (2002) demonstrated that the 48-run plan obtained by this method using Bhattacharya (1944) solution of the BIB design with parameters $v = 16, b = 24, r = 9, k = 6, \lambda = 3$ is not isomorphic to the plan obtained by selecting any 16 columns in a fold-over Hadamard matrix of order 24.

5.6 Box-Behnken Designs

In the response surface methodology, the objective is to determine the levels of the factors to optimize the response. To this end, at the last stage, a second degree polynomial is fitted for the response using the experimental factors as independent variables. The design used to fit the second degree polynomial is called second-order Rotatable Design, if the variance of the estimated response is constant on spherical contours of the factor levels.

Let the rotatable design consist of n design points and t factors. Let $(x_{i1}, x_{i2}, \ldots, x_{it})$ be the levels of the t factors for the ith design point. The following five conditions on the levels of the factors are necessary for the design to be rotatable (see Box and Hunter, 1957; Raghavarao, 1971):

$$\text{A. } \sum_i x_{iu} = 0, \quad \sum_i x_{iu}x_{iu'} = 0, \quad \sum_i x_{iu}x_{iu'}^2 = 0,$$

$$\sum_i x_{iu}^3 = 0, \quad \sum_i x_{iu}x_{iu'}^3 = 0, \quad \sum_i x_{iu}x_{iu'}x_{iu''}^2 = 0,$$

$$\sum_i x_{iu}x_{iu'}x_{iu''} = 0, \quad \sum_i x_{iu}x_{iu'}x_{iu''}x_{iu'''} = 0,$$

$$u, u', u'', u''' = 1, 2, \ldots, t$$

$$u \neq u' \neq u'' \neq u''' \neq u.$$

$$\text{B(i). } \sum_i x_{iu}^2 = n\lambda_2, \quad u = 1, 2, \ldots, t.$$

$$\text{B(ii). } \sum_i x_{iu}^4 = 3n\lambda_4, \quad u = 1, 2, \ldots, t.$$

C. $\sum_i x_{iu}^2 x_{iu'}^2 = \text{constant}, u \neq u', \quad u, u' = 1, 2, \ldots, t.$

D. $\sum_i x_{iu}^4 = 3 \sum_i x_{iu}^2 x_{iu'}^2, \quad u \neq u', \quad u, u' = 1, 2, \ldots, t.$

E. $\lambda_4/\lambda_2^2 > t/(t+2).$

Condition E is needed to estimate all the parameters in the model. The levels will be scaled so that $\lambda_2 = 1$.

Box and Behnken (1960) gave an interesting method of constructing second-order rotatable designs using BIB designs with parameters $v = t, b, r, k, \lambda$ satisfying $r = 3\lambda$. Let N be the incidence matrix of the BIB design. From each column of N we can generate 2^k design points by considering all possible combinations $\pm a$ for the nonzero entries in that column, and determine "a" from the scaling condition $\lambda_2 = 1$. The $2^k b$ points generated in this way with at least one central point of 0 level of all factors is a second-order rotatable design.

Using the BIB design with parameters $v = 7 = b, r = 3 = k, \lambda = 1$, we can construct a second-order rotatable design in 56 non-central points. Taking the solution (4.1), the set (0, 1, 3) will give the following 8 runs:

	Factor						
Run	0	1	2	3	4	5	6
1	a	a	0	a	0	0	0
2	a	a	0	$-a$	0	0	0
3	a	$-a$	0	a	0	0	0
4	a	$-a$	0	$-a$	0	0	0
5	$-a$	a	0	a	0	0	0
6	$-a$	a	0	$-a$	0	0	0
7	$-a$	$-a$	0	a	0	0	0
8	$-a$	$-a$	0	$-a$	0	0	0

The other sets provide the other runs. Taking one central point, $n = 57$, and a can be found from B(i) as

$$24a^2 = 57.$$

Das and Narasimham (1962) extended Box and Behnken results to BIB designs with $r \neq 3\lambda$ and their designs contain more than $b2^k$ design points.

For further details on rotatable designs, see Myers and Montogomery (1995) and Raghavarao (1971).

5.7 Intercropping Experiments

Federer (1993, 1998) discussed extensively intercropping experiments where two or more cultivars are used on the same area of land. If there are m cultivars available in a study, one can form v nonempty sets $S_1, S_2, \ldots, S_v$ of the m cultivators, where the sets may or may not be of equal size and may be of any cardinality $1, 2, \ldots, m$. The sets $S_1, S_2, \ldots, S_v$ are the v treatments and they can be experimented in a completely randomized design, or any block design, or other designs eliminating heterogeneity in several directions.

Let $S_\alpha = \{i_1, i_2, \ldots, i_n\}$ consist of n cultivars $i_1, i_2, \ldots, i_n$ and let for simplicity an orthogonal design is used with the treatments $S_1, S_2, \ldots, S_v$. Let $\bar{Y}_{i_j(S_\alpha)}$ be the mean response on i_j cultivar used in the mixture S_α of cultivars. $E(\bar{Y}_{i_1(S_\alpha)})$, the expected response, for example, of $\bar{Y}_{i_1(S_\alpha)}$ can be written as

$$E(\bar{Y}_{i_1(S_\alpha)}) = \mu + \tau_{i_1}^* + \delta_{i_1} + \sum_{j=2}^{n} \gamma_{i_1(i_j)} + \sum_{\substack{j,j'=2 \\ j<j'}}^{n} \gamma_{i_1(i_j,i_{j'})} + \cdots + \gamma_{i_1(i_2 i_3 \cdots i_n)}, \quad (5.9)$$

where

1. $\mu + \tau_{i_1}^*$ is the performance of ith cultivar as a monoculture or uniblend;
2. δ_{i_1} is the general mixing ability of cultivar i_1 with other cultivars;
3. γ's are specific mixing abilities of different orders. $\gamma_{i_1(i_j)}$ is the first-order specific mixing ability of i_1 cultivar with i_j cultivar. Clearly $\gamma_{i_1(i_j)} \neq \gamma_{i_j(i_1)}$, and the specific mixing abilities are not interactions.

Uni-blends are required to draw conclusions about general mixing abilities, failing which $\tau_{i_1}^*$ and δ_{i_1} will be confounded and inferences can be drawn only on $\tau_{i_1} = \tau_{i_1}^* + \delta_{i_1}$.

Federer and Raghavarao (1987), Raghavarao and Federer (2003), restricted attention to blends using the same number of cultivars, and used BIB designs in intercropping experiments. We will now discuss their results.

The effects can be reparametrized, if necessary, and side conditions on the parameters can be imposed. If the experimenter is interested in drawing inferences upto second-order specific mixing abilities. We assume

$$\sum_{i=1}^{m} \tau_i = 0; \quad \sum_{\substack{j_1=1 \\ j_1 \neq i}}^{m} \gamma_{i(j_1)} = 0, \quad \sum_{\substack{j_2=1 \\ j_1 \neq j_2 \neq i \neq j_1}}^{m} \gamma_{i(j_1,j_2)} = 0. \quad (5.10)$$

The number of independent τ_i parameters are $m - 1$, the number of independent $\gamma_{i(j_1)}$ parameters are $m(m - 2)$ and the number of independent $\gamma_{i(j_1, j_2)}$ parameters are $\binom{m}{2}(m - 4)$. Hence the total number of parameters to be estimated is

$$1 + (m - 1) + m(m - 2) + \binom{m}{2}(m - 4) = m(m - 1)(m - 2)/2. \quad (5.11)$$

The $m(m - 1)(m - 2)/2$ observations can be obtained by considering the v blends $S_1, S_2, \ldots, S_v$ to be the sets of the irreducible BIB designs consisting of sets of 3 cultivars taken from m cultivars and responses are obtained from each cultivar in each of the $\binom{m}{3}$ blends.

The least squares solutions are:

$$\hat{\tau}_i = \binom{m - 1}{2}^{-1} \sum_{\alpha; i \in S_\alpha} \bar{Y}_{i(S_\alpha)} - \tilde{Y},$$

$$\hat{\gamma}_{i(j)} = \frac{1}{(m - 1)(m - 3)} \left\{ (m - 3) \sum_{\alpha; i, j \in S_\alpha} \bar{Y}_{i(S_\alpha)} - 2 \sum_{\substack{\alpha; i \in S_\alpha \\ j \notin S_\alpha}} \bar{Y}_{i(S_\alpha)} \right\},$$

$$\hat{\gamma}_{i(j,k)} = \frac{1}{(m - 2)(m - 3)} \left\{ (m - 3)(m - 4) \sum_{\alpha; i, j, k \in S_\alpha} \bar{Y}_{i(S_\alpha)} - (m - 4) \right.$$

$$\left. \times \sum_{\alpha; i, j \in S_\alpha, k \notin S_\alpha} \bar{Y}_{i(S_\alpha)} - (m - 4) \sum_{\alpha; i, k \in S_\alpha; j \notin S_\alpha} \bar{Y}_{i(S_\alpha)} + 2 \sum_{\alpha; i \in S_\alpha; j, k \notin S_\alpha} \bar{Y}_{i(S_\alpha)} \right\},$$

$$(5.12)$$

where

$$\tilde{Y} = \sum_{i,\alpha} \bar{Y}_{i(S_\alpha)} \bigg/ \left\{ \binom{m}{3} 3 \right\}.$$

These designs are minimal in the sense that we have just enough observations to estimate all the parameters in the model. For further details see Federer and Raghavarao (1987).

Raghavarao and Federer (2003) considered another class of minimal fractional combinatorial treatment designs. Let us be interested in the general mixing ability and first order specific mixing ability of m cultivars. The number of independent parameters to be estimated is $1 + (m - 1) + m(m - 2) = m(m - 1)$ and if we use blocks of size 3, we need $m(m - 1)/3$ blends.

Let $T_1, T_2, \ldots, T_{m-1}$ be pairs of symbols from a set $\{0, 1, 2, \ldots, m - 2\}$ where every symbol occurs exactly twice. Consider a chain, $\theta_0, \theta_1, \theta_2, \ldots, \theta_\ell, \theta_0$, constructed from the elements of the set, where consecutive symbols in the chain occur together in one of the pairs $T_i (i = 1, 2, m - 1)$. The chain is said to be complete if $\{\theta_0, \theta_1, \theta_2, \ldots, \theta_\ell\} = \{0, 1, \ldots, m - 2\}$. Consider the sets

$$(5, 0, 1), (5, 1, 2), (5, 2, 3), (5, 3, 4), (5, 4, 0),$$

$$(0, 1, 3), (1, 2, 4), (2, 3, 0), (3, 4, 1), (4, 0, 2)$$

of a BIB design with parameters $v = 6, b = 10, r = 5, k = 3, \lambda = 2$. Consider the pairs of symbols that occur with 4, say, $T_1 = \{3, 5\}$, $T_2 = \{0, 5\}$, $T_3 = \{1, 2\}$, $T_4 = \{1, 3\}$ and $T_5 = \{0, 2\}$. Using these T_1–T_5, we can form the chain $\theta_0 = 0$, $\theta_1 = 2, \theta_2 = 1, \theta_3 = 3, \theta_4 = 5, \theta_0 = 0$, where the consecutive symbols in the chain belong to the sets T_i. Further $\{\theta_0, \theta_1, \theta_2, \theta_3, \theta_4\} = \{0, 1, 2, 3, 5\}$ and this chain is complete. The following theorem can be easily proved.

Theorem 5.3 (*Raghavarao and Federer*, 2003) *If there exists a BIB design with parameters $v = m$, $b = m(m - 1)/3$, $r = m - 1$, $k = 3$, $\lambda = 2$, for m even, such that for every symbol θ of the BIB design, the chain based on the other pair of symbols occurring in the sets of the design is complete, then the sets of the design form a minimal treatment design for m cultivars in blends of size 3 that are capable of estimating general mixing ability and first-order specific mixing ability contrasts.*

They found that for the same parameters of a BIB design, one solution satisfies the chain condition while the other does not satisfy the chain condition.

5.8 Valuation Studies

Valuation of different factors is very important in decision making in economic and marketing settings. In those problems, factors are called attributes. The attributes are of two types:

1. cost/benefit, and
2. generic.

The first type of attributes, for example, in a car buying setting are price, gas mileage, roominess, reliability, etc. The color of the car is a generic attribute. In this section we consider cost/benefit type of attributes.

If a respondent is asked to choose the best level of attributes by asking about a single attribute at one time, obviously the smallest level of the cost attribute and the highest level of the benefit attribute will be selected. Clearly an item cannot be marketed with such optimal levels of attributes.

It thus becomes necessary to form a choice set consisting of different profiles indicating different level combinations of attributes and asking the respondent to

1. rank all profiles, or
2. choose the best profile, or
3. choose the best profile and score it on a 10-point scale, or
4. assign a given number of points to each profile in the choice set, etc.

If the choice set contains a profile that has smallest cost attributes and highest benefit attributes, that profile is a natural choice because that profile dominates other profiles. On the contrary a profile with highest cost attributes and lowest benefit attributes will never be selected because it is dominated by other profiles. For this reason, dominated or dominating profiles must be excluded from the choice sets.

Let us order the cost attributes from highest to lowest and the benefits attributes from lowest to highest and let there be n attributes in the investigation. We introduce

Definition 5.4 A choice set S is said to be *Pareto Optimal* (PO) if for any two profiles $(u_1, u_2, \ldots, u_n)$, $(w_1, w_2, \ldots, w_n) \in S$, $u_i < w_i$, then there exist a $j \neq i$ such that $u_j > w_j$.

In a Pareto Optimal choice set, no profile dominates or is dominated by other profiles, and it is advisable that the respondents be given a PO choice set for providing the response. By giving PO choice sets, we can find respondent trade-offs between attributes.

Let $Y_{u_1 u_2 \ldots u_n}$ be the response to the profile $(u_1, u_2, \ldots, u_n)$ from a choice set S, which is the rank, or average score, or the logit/probit of the proportion of times the profile is selected. We assume the main effects model

$$Y_{u_1 u_2 \cdots u_n} = \mu + \sum_i \alpha_{u_i}^i + e_{u_1 u_2 \cdots u_n}, \qquad (5.13)$$

where μ is the general mean, $\alpha_{u_i}^i$ is the effect of the u_i level of the ith attribute. Different assumptions about the error terms $e_{u_1 u_2 \cdots u_n}$ can be made; however, to discuss the optimality we assume $e_{u_1 u_2 \cdots u_n}$ to be independently and identically distributed with mean zero and constant variance σ^2. The design is said to be connected main effects plan if we can estimate all main effects contrasts from the model (5.13).

In this section we consider only 2 level attributes, so that we have a 2^n experiment. Clearly, the sets $S_\ell = \{(u_1, u_2, \ldots, u_n) | \sum_i u_i = \ell\}$ are PO subsets. Raghavarao and Wiley (1998) considered a general setting and their results in this particular case implies that a single PO choice set S_ℓ is not a connected main effects plan, and two choice sets $S_{\lfloor n/2 \rfloor}$ and $S_{\lfloor n/2 \rfloor+1}$ is a connected main effects plan, where $\lfloor \bullet \rfloor$ is the greatest integer function. The two choice sets together is not a PO subset. One set of respondents evaluate the profiles in $S_{\lfloor n/2 \rfloor}$, while another set of respondents evaluate the profiles in the other PO choice set and the combined responses will be analyzed for estimating main effects.

Raghavarao and Zhang (2002) showed that any two PO choice sets S_ℓ and $S_{\ell'}$ for $\ell \neq \ell'$ constitute a 2^n experiment connected main effects plan. They considered the information per profile, θ, which is the reciprocal of the average of variances of estimating main effects divided by the number of profiles used, to determine optimal designs and proved the following theorem:

Theorem 5.4 (*Raghavarao and Zhang, 2002*) *For* $n = m^2, \theta = 1$ *for the choice sets* $S_{(n-m)/2}$ *and* $S_{(n+m)/2}$; *for* $n = m(m + 2), \theta = 1$ *for the choice sets* $S_{(n-m)/2}$ *and* $S_{(n+m+2)/2}$; *and in all other cases* $\theta < 1$.

The number of profiles in the optimal choice sets are too many and they are not of practical use. Smaller choice sets having the same $\theta = 1$ can be found using BIB designs.

If a BIB design exists with $v = n, b, r, k, \lambda$ a PO choice set of b profiles with n attributes can be constructed. Each set of the BIB design provides a profile of the choice set with the presence (absence) of the symbol interpreted as high (low) level of the corresponding level. Raghavarao and Zhang (2002) established

Theorem 5.5 1. *The BIB design with parameters* $v = n = m^2, b = m(m + 1)$, $r = (m^2 - 1)/2, k = m(m - 1)/2, \lambda = (m + 1)(m - 2)/4$ *and its complement provide two PO choice sets giving* $\theta = 1$, *when* $m = 4t + 3$ *is a prime or prime power.*
2. *The BIB design with parameters* $v = n = m^2, b = 2m(m + 1), r = (m^2 - 1)$, $k = m(m - 1)/2, \lambda = (m + 1)(m - 2)/2$ *and its complement provide two PO choice sets giving* $\theta = 1$, *when* $m = 4t + 1$ *is a prime or prime power.*

For $n = 9$, the optimal connected main effects plan consists of 2 PO choice sets S_3 and S_6 each with 84 profiles. From Theorem 5.5, we can get an alternative plan using the BIB design with parameters $v = 9, b = 12, r = 4, k = 3, \lambda = 1$ and its complement, and the number of profiles in each of the two choice sets in this plan is only 12.

Let us illustrate the forming of choice sets in a car purchase setting. Nine attributes of interest are:

A: Price ($a_0 = \$25,000$; $a_1 = \$20,000$)

B: Gas mileage ($b_0 = 20\,\text{mpg}$; $b_1 = 25\,\text{mpg}$)

C: Roominess ($c_0 = $ less; $c_1 = $ more)

D: Trunk space ($d_0 = $ less; $d_1 = $ more)

E: Finance charge ($e_0 = 3\%$ for 3 years; $e_1 = 0\%$ for 3 years)

F: First year maintenance cost ($f_0 = \$250$; $f_1 = $ free)

G: Road side assistance ($g_0 = $ nominal charge; $g_1 = $ free)

H: Horse power ($h_0 = 150$; $h_1 = 180$)

I: Side air bags ($i_0 = $ no; $i_1 = $ yes)

Using Theorem 5.5, we form 2 choice sets one using the BIB design with parameters $v = 9, b = 12, r = 4, k = 3, \lambda = 1$ and another using its complement. The profiles of the 2 choice sets are given in Table 5.1.

We will now illustrate the analysis using logit formulation. Let $X_1, X_2, \ldots, X_t$ be multinomial variables with n trials and probabilities $\theta_1, \theta_2, \ldots, \theta_t$ $\left(\sum \theta_i = 1\right)$. Then we know that

$$E(X_i/n) = \theta_i, \quad \text{Var}(X_i/n) = \theta_i(1 - \theta_i)/n, \quad \text{Cov}(X_i/n, X_j/n) = -\theta_i\theta_j/n.$$

For $i = 1, 2, \ldots, t$, let $W_i = \ln[X_i/(n - X_i)]$, be the logit transform of X_i/n. From the delta method,

$$\text{Var}(W_i) = 1/[n\theta_i(1 - \theta_i)], \quad \text{Cov}(W_i, W_j) = -1/[n(1 - \theta_i)(1 - \theta_j)]. \tag{5.14}$$

We will consider attributes A, B, C, D in the car purchasing example discussed earlier and form 2 choice sets. Each choice set will be evaluated by 100 volunteers and each shows their preference. We will include a no choice option in each set to remove the singularity of the dispersion matrix. The data and transformations are summarized in Table 5.2.

Let $\mathbf{W}' = (W_{11}, W_{12}, W_{13}, W_{14}, W_{21}, W_{22}, W_{23}, W_{24})$, and $\boldsymbol{\beta}' = (\mu, A, B, C, D)$, where A, B, C, D are the main effects of the four attributes.

Table 5.1. Choice sets for a 2^9 experiment of car purchasing.

Choice Set	Sets of BIB Design	Profiles
I	a, b, c	$a_1b_1c_1d_0e_0f_0g_0h_0i_0$
	d, e, f	$a_0b_0c_0d_1e_1f_1g_0h_0i_0$
	g, h, i	$a_0b_0c_0d_0e_0f_0g_1h_1i_1$
	a, d, g	$a_1b_0c_0d_1e_0f_0g_1h_0i_0$
	b, e, h	$a_0b_1c_0d_0e_1f_0g_0h_1i_0$
	c, f, i	$a_0b_0c_1d_0e_0f_1g_0h_0i_1$
	a, e, i	$a_1b_0c_0d_0e_1f_0g_0h_0i_1$
	b, f, g	$a_0b_1c_0d_0e_0f_1g_1h_0i_0$
	c, d, h	$a_0b_0c_1d_1e_0f_0g_0h_1i_0$
	a, f, h	$a_1b_0c_0d_0e_0f_1g_0h_1i_0$
	b, d, i	$a_0b_1c_0d_1e_0f_0g_0h_0i_1$
	c, e, g	$a_0b_0c_1d_0e_1f_0g_1h_0i_0$
II	d, e, f, g, h, i	$a_0b_0c_0d_1e_1f_1g_1h_1i_1$
	a, b, c, g, h, i	$a_1b_1c_1d_0e_0f_0g_1h_1i_1$
	a, b, c, d, e, f	$a_1b_1c_1d_1e_1f_1g_0h_0i_0$
	b, c, e, f, h, i	$a_0b_1c_1d_0e_1f_1g_0h_1i_1$
	a, c, d, f, g, i	$a_1b_0c_1d_1e_0f_1g_1h_0i_1$
	a, b, d, e, g, h	$a_1b_1c_0d_1e_1f_0g_1h_1i_0$
	b, c, d, f, g, h	$a_0b_1c_1d_1e_0f_1g_1h_1i_0$
	a, c, d, e, h, i	$a_1b_0c_1d_1e_1f_0g_0h_1i_1$
	a, b, e, f, g, i	$a_1b_1c_0d_0e_1f_1g_1h_0i_1$
	b, c, d, e, g, i	$a_0b_1c_1d_1e_1f_0g_1h_0i_1$
	a, c, e, f, g, h	$a_1b_0c_1d_0e_1f_1g_1h_1i_0$
	a, b, d, f, h, i	$a_1b_1c_0d_1e_0f_1g_0h_1i_1$

Table 5.2. Artificial data in a study.

	Set I			Set II		
Profiles	Proportion Selecting $(X_{1i}/100)$	$W_{1i} = \ln(X_{1i}/(100 - X_{1i}))$		Profiles	Proportion Selecting $(X_{2i}/100)$	$W_{2i} = \ln(X_{2i}/(100 - X_{2i}))$
No Choice	0.1	—		No Choice	0.2	—
$a_0b_1c_1d_1$	0.2	$W_{11} = -1.386$		$a_1b_0c_0d_0$	0.3	$W_{21} = -0.847$
$a_1b_0c_1d_1$	0.2	$W_{12} = -1.386$		$a_0b_1c_0d_0$	0.3	$W_{22} = -0.847$
$a_1b_1c_0d_1$	0.25	$W_{13} = -1.099$		$a_0b_0c_1d_0$	0.1	$W_{23} = -2.197$
$a_1b_1c_1d_0$	0.25	$W_{14} = -1.099$		$a_0b_0c_0d_1$	0.1	$W_{24} = -2.197$

Further, let

$$
X = \begin{pmatrix}
1 & -1 & 1 & 1 & 1 \\
1 & 1 & -1 & 1 & 1 \\
1 & 1 & 1 & -1 & 1 \\
1 & 1 & 1 & 1 & -1 \\
1 & 1 & -1 & -1 & -1 \\
1 & -1 & 1 & -1 & -1 \\
1 & -1 & -1 & 1 & -1 \\
1 & -1 & -1 & -1 & 1
\end{pmatrix},
$$

$$
V = \begin{pmatrix}
0.063 & -0.016 & -0.017 & -0.017 & 0 & 0 & 0 & 0 \\
-0.016 & 0.063 & -0.017 & -0.017 & 0 & 0 & 0 & 0 \\
-0.017 & -0.017 & 0.053 & -0.018 & 0 & 0 & 0 & 0 \\
-0.017 & -0.017 & -0.018 & 0.053 & 0 & 0 & 0 & 0 \\
0 & 0 & 0 & 0 & 0.048 & -0.020 & -0.016 & -0.016 \\
0 & 0 & 0 & 0 & -0.020 & 0.048 & -0.016 & -0.016 \\
0 & 0 & 0 & 0 & -0.016 & -0.016 & 0.111 & -0.012 \\
0 & 0 & 0 & 0 & -0.016 & -0.016 & -0.012 & 0.111
\end{pmatrix},
$$

V is calculated using the estimated variances and covariances of W_{ij}. Then $E(\mathbf{W}) = X\boldsymbol{\beta}$ and Var($\mathbf{W}$) is approximately V. From the weighted normal equations, we get

$$
\hat{\boldsymbol{\beta}}' = \mathbf{W}'V^{-1}X(X'V^{-1}X)^{-1} = (-1.34 \quad 0.23 \quad 0.23 \quad -0.15 \quad -0.15),
$$

$$
\hat{\text{Var}}\left(\hat{\boldsymbol{\beta}}\right) = \left(X'V^{-1}X\right)^{-1} = \begin{pmatrix}
0.0021 & -0.0017 & -0.0017 & 0.0006 & 0.0006 \\
-0.0017 & 0.0084 & -0.0007 & -0.0022 & -0.0022 \\
-0.0017 & -0.0007 & 0.0084 & -0.0022 & -0.0022 \\
0.0006 & -0.0022 & -0.0022 & 0.0079 & -0.0033 \\
0.0006 & -0.0022 & -0.0022 & -0.0033 & 0.0079
\end{pmatrix}.
$$

Different parameter estimates, test statistics and the p-values are summarized in Table 5.3.

The lack of fit is not significant at 0.05 level and the main effects of A and B are significant at 0.05 level.

Some researchers use ln(proportion selecting a profile/proportion selecting no choice) as the response variable, and using appropriate variances and covariances for that variable, analyze the data by the method of weighted least squares.

Table 5.3. Parameter estimates and tests.

	Est.	Std. Error	Test Statistic χ^2	df	p-value
Lack of fit	—	—	$\mathbf{W'}V^{-1}\mathbf{W} - \mathbf{W'}V^{-1}X\hat{\beta}$ $= 1302.959 - 1296.332$ $= 6.627$	$8 - 5$ $= 3$	0.085
A	0.23	0.092	6.25	1	0.012
B	0.23	0.092	6.25	1	0.012
C	-0.15	0.089	2.84	1	0.092
D	-0.15	0.089	2.84	1	0.092

5.9 Tournament and Lotto Designs

Let $4t + 3$ players (or teams) participate in a tournament. There are $2t + 1$ courts (or tables) available for the matches and the players (or teams) play several rounds of the game to decide the winner. We want to arrange the tournament schedule in the following way:

1. There are $4t + 3$ rounds of the game and in each round one player (or team) will not play.
2. Every player (or team) plays in $4t + 2$ rounds, once on each side of each court (or table).
3. Every player (or team) plays opposite to other player (or team) exactly once.
4. Every pair of distinct players (or teams) plays on the same side of court (or table) equal number of times.

The solution is based on BIB designs. Let us assume that $4t + 3$ is a prime or prime power, and let $\alpha_0 = 0$, $\alpha_i = x^i$, $i = 1, \ldots, 4t + 2$ be the elements of $GF(4t + 3)$, where x is a primitive root of the field. Note that $x^{4t+2} = x^0 = 1$. We denote the players (or teams) by α_i for $i = 0, 1, \ldots, 4t + 2$. Let a pair (x_i, y_i) represent the players (or teams) playing in the ith court in a round, where $x_i, y_i = \alpha_0, \alpha_1, \ldots, \alpha_{4t+2}$ and $x_i \neq y_i$. If we can determine the first round as $S = \{(x_1, y_1), (x_2, y_2), \ldots, (x_{2t+1}, y_{2t+1})\}$ with distinct symbols such that

1. among the $4t(2t + 1)$ differences $\pm(x_i - x_j)$, $\pm(y_i - y_j)$, for $i, j = 1, 2, \ldots, 2t + 1$; $i \neq j$, all the nonzero elements of $GF(4t + 3)$ occur $2t$ times; and

2. among the $4t + 2$ differences, $\pm(x_i - y_i), i = 1, 2, \ldots, 2t + 1$, each nonzero element of GF$(4t + 3)$ occurs exactly once,

then the $4t + 3$ rounds are given by

$$S_\theta = \{(x_1 + \theta, y_1 + \theta), (x_2 + \theta, y_2 + \theta), \ldots, (x_{2t+1} + \theta, y_{2t+1} + \theta)\},$$

$\theta \in$ GF$(4t + 3)$, where $x_i + \theta, y_i + \theta \in$ GF$(4t + 3)$. This result follows from Theorem 4.15.

Clearly $S = \{(x^0, x), (x^2, x^3), \ldots, (x^{4t}, x^{4t+1})\}$, satisfies the requirements of the first round. With 7 players (or teams) the 7 rounds are

$$S_0 = \{(1, 3), (2, 6), (4, 5)\},$$
$$S_1 = \{(2, 4), (3, 0), (5, 6)\},$$
$$S_2 = \{(3, 5), (4, 1), (6, 0)\},$$
$$S_3 = \{(4, 6), (5, 2), (0, 1)\},$$
$$S_4 = \{5, 0), (6, 3), (1, 2)\},$$
$$S_5 = \{(6, 1), (0, 4), (2, 3)\},$$
$$S_6 = \{(0, 2), (1, 5), (3, 4)\}.$$

Schellenberg, Van Rees and Vanstone (1977) considered balanced tournament designs using BIB designs and other combinatorial structures. Also, see the excellent book of Anderson (1997) for the use of block designs as tournament designs. Nested BIB designs are useful to arrange Bridge and other tournaments and we will return to this topic in Sec. 10.4.

In most of the lotteries, the ticket purchasers are asked to select k numbers from given n numbers for a ticket. Any number of tickets can be purchased by an individual. The house closes the selling of tickets and draws p numbers randomly from the given n numbers from which the tickets are purchased. A purchased ticket is a winning ticket if among the k numbers of that ticket at least t of them are in the p numbers selected by the house. The problem is to form minimum number b of tickets so that at least one of the b tickets is a winning ticket.

Theorem 5.6 (*Li and Van Rees*, 2000) *The sets of a BIB design with parameters* $v = n, b, r, k, \lambda$ *is a winning ticket, if*

$$\left\lfloor \frac{pr}{t-1} \right\rfloor \binom{t-1}{2} + \left(\frac{pr - \left\lfloor \frac{pr}{t-1} \right\rfloor (t-1)}{2} \right) < \binom{p}{2} \lambda, \qquad (5.15)$$

where $\lfloor \bullet \rfloor$ *is the greatest integer function.*

Proof. Let P be a set of p symbols and no subset of P of t or more symbols is contained in the sets of the BIB design. The sets of BIB design contain $\binom{p}{2}\lambda$ pairs of the symbols of P. Because of the specification of P, the maximum number of pairs of P forced in the sets of BIB design is the left-hand side of (5.14), and hence a contradiction.

For further details on Lotto designs the reader is referred to Li and Van Rees (2000).

5.10 Balanced Half-Samples

Consider a stratified sampling setting with L strata where the hth stratum has weight $W_h = N_h/N$, N_h and N being the hth stratum and population sizes, respectively. Let a sample of size $2L$ be drawn from that population with assignment $n_h = 2$ for the hth stratum.

Let t be the smallest positive integer such that $4t - 1 \geq L$, and let N be the incidence matrix of the Hadamard BIB design with parameters $v = 4t - 1 = b$, $r = 2t - 1 = k, \lambda = t - 1$. Let

$$M = [N | \mathbf{1}_{4t-1}].$$

Choose any L rows of M to constitute a matrix M_1 of order $L \times 4t$. The rows of M_1 are identified with the L strata and the 1 (0) elements of the ith row are arbitrarily identified with the 2 sample observations from the ith stratum. The $4t$ columns of M_1 are the $4t$ half-samples.

From the jth column, we form an estimator of population mean $\bar{Y}$ as

$$\bar{Y}_{st(j)} = \sum_{i=1}^{L} W_i y_{ij},$$

where y_{ij} is the observation corresponding to 1 (0) element of the ith row (stratum) in the jth column. Combining all the half-samples estimators, we get the estimator of $\bar{Y}$ as

$$\hat{\bar{Y}} = \sum_{j=1}^{4t} \bar{Y}_{st(j)}/4t,$$

with the estimated variance

$$\hat{\text{Var}}\left(\hat{\bar{Y}}\right) = \sum_{j=1}^{4t} \left(\bar{Y}_{st(j)} - \hat{\bar{Y}}\right)^2 \Big/ 4t.$$

From the properties of BIB designs it can be verified that each of the two observations from each stratum occurs in $2t$ half samples, and each of the 4 pairs of observations from any two strata occurs in t half samples. The results of this section are originally due to McCarthy (1969).

6

t-Designs

6.1 Introduction

We define a t-design in Definition 6.1.

Definition 6.1 An arrangement of v symbols in b sets of size k is said to be a *t-design* if

(1) every symbol occurs at most once in a set,
(2) every symbol occurs in r sets,
(3) every t symbols occur together in λ_t sets.

The following 30 sets in 10 symbols $0, 1, \ldots, 9$ is a 3-design with parameters $v = 10, b = 30, r = 12, k = 4, \lambda_3 = 1$:

$$
\begin{aligned}
&(0, 1, 2, 6);\ (0, 1, 3, 4);\ (0, 2, 5, 8);\\
&(1, 2, 3, 7);\ (1, 2, 4, 5),\ (1, 3, 6, 9),\\
&(2, 3, 4, 8);\ (2, 3, 5, 6);\ (2, 4, 7, 0);\\
&(3, 4, 5, 9);\ (3, 4, 6, 7);\ (3, 5, 8, 1);\\
&(4, 5, 6, 0);\ (4, 5, 7, 8);\ (4, 6, 9, 2);\\
&(5, 6, 7, 1);\ (5, 6, 8, 9);\ \ (5, 7, 0, 3)\\
&(6, 7, 8, 2);\ (6, 7, 9, 0);\ (6, 8, 1, 4);\\
&(7, 8, 9, 3);\ (7, 8, 0, 1);\ (7, 9, 2, 5);\\
&(8, 9, 0, 4);\ (8, 9, 1, 2);\ (8, 0, 3, 6);\\
&(9, 0, 1, 5);\ (9, 0, 2, 3);\ (9, 1, 4, 7).
\end{aligned}
\tag{6.1}
$$

A 2-design is the BIB design discussed in Chaps. 4 and 5. A 3-design is also known as a doubly balanced incomplete block design and Calvin (1954) used it in experiments. In the last section of this chapter, we will discuss the use of 3-designs as brand availability designs. Excellent source references to t-designs are the review papers by Hedayat and Kageyama (1980), Kageyama and Hedayat (1983), and Kreher (1996). In this chapter we will discuss some of the statistical applications and combinatorics of 3-designs.

Trivially all $\binom{v}{k}$ combinations of k symbols from the v symbols, where $k \geq t$ is a t-design, and we call it an irreducible t-design. A t-design is also an s-design for $s < t$. In fact, given a t- design, consider the sets in which s given symbols occur, say, λ_s. Enumerating in two ways of enlarging the s symbols to given t symbols, we get

$$\lambda_s \binom{k-s}{t-s} = \lambda_t \binom{v-s}{t-s},$$

and this is not depending on the selected s symbols. Thus a t-design, for $s < t$, is an s-design with

$$\lambda_s = \lambda_t \binom{v-s}{t-s} \Big/ \binom{k-s}{t-s}. \tag{6.2}$$

A t-design with $\lambda_t = 1$ is called a Steiner system and is denoted by $S(t, k, v)$. While infinitely many Steiner systems exist for $t \leq 3$, very few exist for $t \geq 4$. Witt systems $S(4, 5, 11)$, $S(5, 6, 12)$, $S(4, 7, 23)$ and $S(5, 8, 24)$ associated with the Mathieu groups and the systems $S(5, 7, 28)$, $S(5, 6, 24)$, $S(5, 6, 48)$, $S(5, 6, 24)$, $S(4, 6, 26)$, $S(4, 5, 23)$, $S(4, 5, 47)$ and $S(4, 5, 83)$ found by Denniston (1976) are some Steiner systems for $t = 4$ or 5.

6.2 Inequalities on b

Raghavarao (1971) using some generalization of incidence matrix proved.

Theorem 6.1 *For a t-design with $v > k + 1$, we have*

$$b \geq (t-1)(v-t+2). \tag{6.3}$$

A t-design can be split into 2 designs. By considering all the sets in which a symbol θ occurs and omitting θ, we get a $(t-1)$-design, while all the sets in which θ does not occur will also be a $(t-1)$- design. Both designs have $v-1$ symbols and the set sizes are $k-1$ and k, respectively. Using this fact repeatedly and Fisher's inequality $b \geq v$ for a BIB design, Dey and Saha (1974) proved

Theorem 6.2 *For a t-design with $v \geq k + t - 1$, we have*

$$b \geq 2^{t-2}(v-t+2). \tag{6.4}$$

By considering the distribution of all s-tuples in the b sets, Ray–Chaudhuri and Wilson (1975) proved

Theorem 6.3 *For a t-design with $t = 2s$, $v \geq k + s$, we have*

$$b \geq \binom{v}{s},\tag{6.5}$$

and with $t = 2s + 1$, $v \geq k + s + 1$, we have

$$b \geq 2\binom{v-1}{s}.\tag{6.6}$$

All the above 3 theorems give the bound $b \geq 2\,(v - 1)$ for 3-designs and this bound is attained for only one series of 3-designs with parameters

$$v = 4t, \quad b = 2(4t - 1), \quad r = 4t - 1, \quad k = 2t, \quad \lambda_3 = t - 1.\tag{6.7}$$

The series (6.7) is conjectured to exist for all t from Hadamard design with parameters $v = 4t - 1 = b$, $r = 2t - 1 = k$, $\lambda = t - 1$ augmented by a new symbol in each set and its complement.

Wilson (1982) gave an alternative proof of Theorem 6.3. Raghavarao, Shrikhande, and Shrikhande (2002) showed that Theorem 6.3 can be established in special cases by elementary means and using appropriate incidence matrices. We will sketch their proof for a 4-design.

Consider the incidence matrix $N_1 = (n_{ij,\ell})$, where for $1 \leq i < j \leq v$, we put $n_{ij,\ell} = 1(0)$, according to the pair of symbols i and j occur together (not) in the ℓth set. N_1 is a $\binom{v}{2} \times b$ matrix. It can be verified that

$$N_1 N_1' = \lambda_2 I_{\binom{v}{2}} + \lambda_3 B_1 + \lambda_4 B_2,\tag{6.8}$$

where B_1 and B_2 are association matrices of triangular association scheme to be defined in Chap. 8. The eigenvalues of $N_1 N_1'$ are positive because $v \geq k + 2$ (see Chap. 8, p. 127) and hence

$$\binom{v}{2} = \mathrm{Rank}(N_1 N_1') = \mathrm{Rank}(N_1) \leq b,$$

establishing the result of Theorem 6.3.

A t-design is said to be tight if the equality specified in Theorem 6.3 is attained. Tight t-designs are discussed by Carmony (1978), Deza (1975), Ito (1975) and Peterson (1977).

6.3 Resistance of Variance Balance for the Loss of a Treatment

Suppose an experiment is planned with v treatments using a variance balanced equi-replicate and equi-block sized incomplete block design, i.e. a BIB design.

During the course of the experiment, the experimenter observes that one of the v treatments is not performing well as anticipated and wants to drop it out from further investigation. He/she also wants that the remaining design in $v - 1$ treatments must be optimal so that it must be variance balanced. If treatment θ of the v treatments is deleted and if treatments i and j occur together with θ in x sets, the variance balance property of the block design in $v - 1$ treatments imply

$$\frac{x}{k-1} + \frac{\lambda - x}{k} = \text{constant},$$

thereby implying that x is the same for every treatment pair i, j. Thus all the triplets of symbols θ, i and j occur equally often and the design is a 3-design.

Using 3-designs in experiments enable the experimenter to miss the responses on any one treatment and still retain the optimality for the remaining part of the experiment. This aspect of 3-designs is discussed by Hedayat and John (1974), P.W.M. John (1976a) and Most (1976).

6.4 Constructions

Theorem 4.14 can be used to establish

Theorem 6.4 *A BIB design with parameters $v = 2k, b', r', k \, (> 2), \lambda'$ and its complement, constitute a 3-design with parameters*

$$v = 2k, \quad b = 2b', \quad r = 2r' = b', \quad k, \quad \lambda_3 = b' - 3r' + 3\lambda'. \tag{6.9}$$

Using the BIB designs with $v = 2k$ given by Preece (1967b) and Theorem 6.4, we can construct many 3-designs.

Theorem 6.4 can be generalized to

Theorem 6.5 *(Saha, 1975a) The sets of a t-design with $v = 2k$, $k > t$ for t even and the complements of the sets of the t design, form a $(t + 1)$-design.*

Proof. Let $\lambda_0 = b, \lambda_1, \lambda_2, \ldots, \lambda_t$ be the λ-parameters of the original t-design. The new design can be verified to be a $(t + 1)$-design with λ-parameters,

$$\begin{aligned}
\lambda_j^* &= \sum_{i=0}^{j} (-1)^i \binom{j}{i} \lambda_i + \lambda_j, \quad j = 0, 1, 2, \ldots, t \\
\lambda_{t+1}^* &= \sum_{i=0}^{t} (-1)^i \binom{t}{i} \lambda_i,
\end{aligned} \tag{6.10}$$

by using the method of inclusion and exclusion.

In EG(n, 2) geometry, by considering the points as symbols and d-flats as sets for $n > d \geq 3$, we get a 3-design with the following parameters (see Kageyama, 1973c):

$$v = 2^n, \quad b = 2^{n-d}\phi(n-1, d-1, 2), \quad r = \phi(n-1, d-1, 2), \quad k = 2^d,$$
$$\lambda_2 = \phi(n-2, d-2, 2), \quad \lambda_3 = \phi(n-3, d-3, 3). \tag{6.11}$$

We have

Theorem 6.6 (*Takahasi*, 1975) *A 3-design with parameters*

$$v = s^{w+1} + 1, \quad b = s^w(s^{n+1} - 1)/(s^2 - 1),$$
$$r = s^w(s^{w+1} - 1)/(s - 1), \quad k = s + 1, \tag{6.12}$$
$$\lambda_2 = (s^{w+1} - 1)/(s - 1), \quad \lambda_3 = 1$$

exists, where s is a prime or prime power, n is odd and $w = (n-1)/2$.

Proof. Consider PG(n, s) geometry for odd n. Let $w = (n-1)/2$. The set of w-flats $\{T_0, T_1, \ldots, T_{v-1}\}$, for $v = (s^{n+1} - 1)/(s^{w+1} - 1) = s^{w+1} + 1$ is said to be a w-spread in PG (n, s) if $T_i \cap T_j = \phi$ for $i, j = 0, 1, \ldots, v - 1; i \neq j$, and $\bigcup_{i=0}^{v-1} T_i$ is the set of all points in PG(n, s). Consider all the lines L_i of the geometry which are not totally contained in a T_j and form sets

$$S_i = \{j | T_j \cap L_i \neq \phi\}.$$

We note that the resulting design will have repeated sets and the distinct sets form the required 3-design.

Raghavarao and Zhou (1997) extended the method of differences of constructing BIB designs to construct 3-designs and we will now discuss their method.

Consider a module $M = \{0, 1, \ldots, v - 1\}$ of v elements. For $i = 1, 2, \ldots, \lfloor(v - 1)/3\rfloor$, where $\lfloor \bullet \rfloor$ is the greatest integer function, the sets of triples $(0, i, 2i), (0, i, 2i + 1), \ldots, (0, i, v - i - 1)$ will be called the set of initial triples. The v sets $(\theta, \theta + i, \theta + j)$ for $\theta \in M$ are the triples developed mod v from the initial triple $(0, i, j)$. They established that every triple of distinct symbols occurs exactly once, where the set of initial triples are developed mod v, whenever $v \not\equiv 0$ (mod 3), and consequently proved.

Theorem 6.7 (*Raghavarao and Zhou*, 1997) *Let $v \not\equiv 0$ (mod 3) and let there exist h sets $S_1, S_2, \ldots, S_h$, each of k distinct elements of the module $M = \{0, 1, \ldots, v-1\}$, such that among the $hk(k - 1)(k - 2)$ distinct triples (ℓ, m, n) where $\ell < m < n$, formed from each of the sets $S_1, S_2, \ldots, S_h$, when written as $(0, m - \ell, n - \ell)$, each of the initial sets of triples developed mod v associated with these triples occurs*

λ_3 *times. Then the vh sets $S_{i\theta}$ for $\theta \in M, i = 1, 2, \ldots, h$, where $S_{i\theta} = S_i + \theta$ form a 3-design with parameters, $v, b = vh, k, r = hk, \lambda_2 = \lambda_3(v - 2)/(k - 2)$, and λ_3.*

We will illustrate Theorem 6.7 in the construction of a 3-design with parameters $v = 17, b = 68, r = 20, k = 5, \lambda_2 = 5, \lambda_3 = 1$.

Consider $M = \{0, 1, 2, \ldots, 16\}$ and consider the 4 sets $S_1 = (0, 1, 2, 8, 11)$; $S_2 = (0, 1, 3, 5, 6)$; $S_3 = (0, 1, 4, 9, 14)$; and $S_4 = (0, 2, 6, 10, 12)$. The triples from S_1, S_2, S_3 and S_4 and the corresponding initial triples are given below:

Triples from S_1, S_2, S_3, S_4	Corresponding initial triples	Triples from S_1, S_2, S_3, S_4	Corresponding initial triples
0, 1, 2	0, 1, 2	0, 1, 4	0, 1, 4
0, 1, 8	0, 1, 8	0, 1, 9	0, 1, 9
0, 1, 11	0, 1, 11	0, 1, 14	0, 1, 14
0, 2, 8	0, 2, 8	0, 4, 9	0, 4, 9
0, 2, 11	0, 2, 11	0, 4, 14	0, 3, 7
0, 8, 11	0, 3, 9	0, 9, 14	0, 3, 12
1, 2, 8	0, 1, 7	1, 4, 9	0, 3, 8
1, 2, 11	0, 1, 10	1, 4, 14	0, 3, 13
1, 8, 11	0, 3, 10	1, 9, 14	0, 4, 12
2, 8, 11	0, 3, 11	4, 9, 14	0, 5, 10
0, 1, 3	0, 1, 3	0, 2, 6	0, 2, 6
0, 1, 5	0, 1, 5	0, 2, 10	0, 2, 10
0, 1, 6	0, 1, 6	0, 2, 12	0, 2, 12
0, 3, 5	0, 2, 14	0, 6, 10	0, 4, 11
0, 3, 6	0, 3, 6	0, 6, 12	0, 5, 11
0, 5, 6	0, 1, 12	0, 10, 12	0, 2, 7
1, 3, 5	0, 2, 4	2, 6, 10	0, 4, 8
1, 3, 6	0, 2, 5	2, 6, 12	0, 4, 10
1, 5, 6	0, 1, 13	2, 10, 12	0, 2, 9
3, 5, 6	0, 1, 15	6, 10, 12	0, 2, 13

Each of the initial triples occurs once in S_1, S_2, S_3, S_4 and hence the 68 sets $S_{i\theta}$ for $i = 1, 2, 3, 4$ and $\theta \in M$ constitute the solution of the 3-design.

Selected references dealing with other construction methods of t-designs are Alltop (1969, 1972), Assmus and Mattson (1969), Hanani (1979), Kageyama (1973c), Kramer (1975), Mills (1978), Noda (1978) and Shrikhande (1973).

6.5 A Cross-Effects Model

Let v brands of a product be available and a consumer has to choose between k brands of the v available brands. The set of k available brands is called a choice set. In market research, volunteers are provided with choice sets to mimic the actual market setting and conclusions are drawn. The volunteer may be asked to give the response in any of the following ways:

1. choose the best brand in the choice set
2. rank the brands of the choice set
3. select the best brand and give a score on an appropriate point scale, etc.

With response 1, the proportion of volunteers selecting each brand will be calculated and it will be transformed by a logit or probit transformation. In the case of second type response, the ranks will be converted to some scores to represent the revenue or total sales of the product. In this way a response or transformed response can be generated for each brand in all created choice sets. The response of a brand in a choice set can be modeled as given in Eq. (4.56). The independence and variance structure of the error terms in the model depends on the type of response used. For getting optimal designs we assume that the errors are identically and independently distributed with mean zero and variance σ^2.

There are two types of problems of interest:

I. Suppose A and B are two brands on the shelf available to the customer and brand C is not available. We are interested to see the effect on the sales of A, when brand B is replaced by brand C. For this purpose, we need to estimate all the cross effects contrasts,

$$\beta_{i(j)} - \beta_{i(j')} \quad \text{for } i \neq j \neq j' \neq i.$$

II. A new shop owner has shelf space to display k brands. He/she wants to find the k brands to be displayed to maximize the revenue by selling that product. Here we need to estimate β_i, brand effects and $\beta_{i(j)}$ brand cross effects individually.

In case I, we assume the model,

$$E(Y_i) = \mu + \beta_i + \sum_{\substack{j \in S \\ j \neq i}} \beta_{i(j)}, \tag{6.13}$$

where S is the choice set, with side conditions

$$\sum_{i=1}^{v} \beta_i = 0, \quad \sum_{\substack{j=1 \\ j \neq i}}^{v} \beta_{i(j)} = 0. \tag{6.14}$$

By having the side conditions (6.14), β_i is not exactly the brand effect; but is the brand effect plus the average of cross effects of other brands on the ith brand.

In case II, we assume

$$E(Y_i) = \beta_i + \sum_{\substack{j \in S \\ j \neq i}} \beta_{i(j)}, \tag{6.15}$$

with no side conditions on the parameters.

Using the model (6.13), Bhaumik (1995) showed that a 3-design is optimal for estimating all contrasts of brand cross effects and brand effects. Using a 3-design with parameters $v, b, r, k, \lambda_2, \lambda_3$, let the responses be obtained for an experiment. Let T_i be the total response for brand i and $T_{i(j)}$ be the response for brand i from the choice sets where brand j is also available. It can easily be verified that

$$\hat{\beta}_i = \frac{1}{r} T_i - \frac{G}{vr},$$
$$\hat{\beta}_{i(j)} = \frac{1}{\lambda_2 - \lambda_3} T_{i(j)} - \frac{\lambda_2}{\lambda_2 - \lambda_3} \frac{T_i}{r}, \tag{6.16}$$

for $j = 1, 2, \ldots, v$; $i = 1, 2, \ldots, v$; $j \neq i$, where $G = \sum_{i=1}^{v} T_i$. The ANOVA table is given in Table 6.1. If the design is saturated and no error df are available, one needs to take multiples of the design and modify Table 6.1 accordingly.

Raghavarao and Wiley (1986) conducted a choice experiment on 8 brands of soft drinks: Coke (C), Diet Coke (DC), Pepsi (P), Diet Pepsi (DP), Seven-up (7). Diet Seven-up (D7), Sprite (S), Diet Sprite (DS). They used the 3-design with parameters $v = 8$, $b = 14$, $r = 7$, $k = 4$, $\lambda_2 = 3$, $\lambda_3 = 1$ and administered the experiment on 112 student volunteers. In their analysis, the cross

Table 6.1. ANOVA for a cross effects model.

Source	df	SS	MS	F
Brand effects	$v - 1$	$\sum_{i=1}^{v} \dfrac{T_i^2}{r} - \dfrac{G^2}{vr}$	MS_b	MS_b/MS_e
Cross effects	$v(v - 2)$	$\sum_{\substack{i,j=1 \\ i \neq j}}^{v} \hat{\beta}_{i(j)} T_{i(j)}$	MS_c	MS_c/MS_e
Error	$v(r - v + 1)$	by subtraction	MS_e	
Total	$vr - 1$	$\sum_{i,j} Y_{ij}^2 - \dfrac{G^2}{vr}$		

effects are significant. The multiple comparisons of cross effects, for example, on DC are

Favorable						Unfavorable
D7	S	7	DP	P	C	DS

The above configuration implies that if a store has DC, S and two other brands except DS, and if S is replaced by DS, the sales of DC will go down. This is expected because diet drinkers will choose and try another diet drink, when available. On the contrary if DS is available and S is not available, and if DS is replaced by S, the sales of DC will go up.

Raghavarao and Zhou (1998) noted that for the model (6.15), the individual parameters β_i and $\beta_{i(j)}$ are non-estimable if the choice sets are all of equal size. They showed that the optimal design for the model (6.15) is a design with unequal set sizes in which every pair of symbols occur together in λ_2 sets and every triple of symbols occur together in λ_3 sets. They gave a list of parameters with two set sizes and $v \leq 10, b \leq 50$.

7

Linked Block Designs

7.1 Dual Designs — Linked Block Designs

If D is a block design with v symbols in b sets having incidence matrix N, the dual design D, given by D^*, is obtained by interchanging the roles of symbols and sets. D^* will have b symbols arranged in v sets with incidence matrix N'.

We noted in Chap. 4 that every pair of distinct sets in a symmetric BIB design with parameters $v = b, r = k, \lambda$, has λ common symbols. Hence the dual of a symmetric BIB design is also a symmetric BIB design with the same set of parameters.

Let us consider a BIB design with $\lambda = 1$. The number of common symbols between any two sets of such a BIB design is at most one. It is easy to note that given a set S_1, the number of sets of the design having one common symbol with S_1 is $n_1 = k(r - 1)$, and all other $n_2 = b - 1 - k(r - 1)$ sets have no symbol in common with S_1. Given two sets S_1 and S_2 having one symbol in common, the number of sets having one symbol in common with each of the sets S_1 and S_2 is $r - 2 + (k - 1)^2 = p_{11}^1$, say. Thus in the dual of the BIB design with $\lambda = 1$, the b symbols occur in v sets such that with every symbol, each of n_1 symbols occur exactly once in the sets, while each of the other n_2 symbols does not occur with it at all. Furthermore, if two symbols occur together in a set, the number of symbols occurring with each of the symbols in the sets of the dual design is p_{11}^1. Such designs are called Partially Balanced Incomplete Block (PBIB) designs with two associate classes and we will study them in detail in the next chapter.

Consider an affine α-resolvable BIB design where the b sets are grouped into t classes each of β sets, and each of the v symbols occur α times in each class. It is known that any two sets of the same class have $k + \lambda - r$ symbols in common, and any two sets of different classes have k^2/v symbols in common. Thus in its dual design, the b symbols will be divided into t groups of β symbols, and symbols of the same group occur together in $k + \lambda - r$ sets, while symbols from different groups occur together in k^2/v sets. Such designs are Group Divisible (GD) designs and form a particular class of PBIB designs.

For other interesting results on dual designs, the reader is referred to Raghavarao (1971, Chap. 10). In the rest of this chapter, we will discuss Linked Block (LB) designs originally introduced by Youden (1951) and studied by Roy and Laha (1956, 1957) which are duals of BIB designs. Formally we define:

Definition 7.1 A *LB design* is an arrangements of v symbols in b sets of size k such that every symbol occurs in r sets and every two distinct sets have μ symbols in common.

v, b, r, k and μ are parameters of LB design, and clearly they satisfy

$$vr = kb, \quad \mu(b-1) = k(r-1). \tag{7.1}$$

If N is the incidence matrix of a LB design, then

$$N'N = (k-\mu)I_b + \mu J_b; \tag{7.2}$$

and $N'N$ is non-singular. Consequently,

$$b \leq v. \tag{7.3}$$

Since the nonzero eigenvalues of NN' and $N'N$ are the same, and as the nonzero eigenvalues of $N'N$ given in (7.2) are $\theta_0 = rk, \theta_1 = k - \mu$ with multiplicities $\alpha_0 = 1, \alpha_1 = b - 1$, we have

Theorem 7.1 *An incomplete block design is a Linked Block Design if NN' has only two nonzero eigenvalues $\theta_0 = rk(>0), \theta_1 = k - \mu(>0)$ with respective multiplicities $\alpha_0 = 1$, and $\alpha_1 = b - 1$, where N is the incidence matrix of the incomplete block design.*

If ℓ is a $b \times 1$ vector such that $\ell' \mathbf{1}_b = 0$ and $\ell'\ell = 1$, satisfying

$$N'N\ell = \theta_1\ell,$$

then

$$NN'\left(\frac{1}{\sqrt{\theta_1}}N\ell\right) = \theta_1\left(\frac{1}{\sqrt{\theta_1}}N\ell\right). \tag{7.4}$$

Thus $\frac{1}{\sqrt{\theta_1}}N\ell$ is a normalized eigenvector corresponding to the nonzero eigenvalue θ_1 of NN'.

The authors feel that LB designs were not given due recognition in the design literature, though their optimality is well established by Shah, Raghavarao and Khatri (1976), and they need fewer blocks than treatments.

7.2 Intra-Block Analysis

With the notation of Chap. 2,

$$C_{\beta|\tau} = kI_b - \frac{1}{r}N'N = \frac{\mu b}{r}I_b - \frac{\mu}{r}J_b,$$

$$\mathbf{Q}_{\beta|\tau} = \mathbf{B} - \frac{1}{r}N'\mathbf{T},$$

and hence

$$\hat{\beta} = \frac{r}{\mu b}\mathbf{Q}_{\beta|\tau}.$$

Thus

$$SS_{B|Tr} = \frac{r}{\mu b}Q'_{\beta|\tau}\mathbf{Q}_{\beta|\tau}. \tag{7.5}$$

$SS_{Tr|B}$ can be obtained as

$$SS_{Tr|B} = SS_{Tr} + SS_{B|Tr} - SS_B. \tag{7.6}$$

The ANOVA Table 2.1 and Type III SS Table 2.3 can be easily set and the equality of treatment effects tested. If inferences on treatment contrasts are needed, $C_{\tau|\beta}$ and its g-inverse can be computed to estimate $\hat{\tau}$ as discussed in Chap. 2.

7.3 Optimality

The following theorem establishes the A-, D-, and E-optimality of LB designs.

Theorem 7.2 (*K. R. Shah, Raghavarao and Khatri, 1976*) *If the class of incomplete equi-block sized, equi-replicated designs with parameters v, b, r, k contains a Linked Block (LB) design, then that design is A-, D- and E-optimal for the estimation of the treatment effects.*

Proof. Let N be the incidence matrix of the block design and consider the normal equations for estimating the parameters μ, β and τ given in Sec. 2.2. Let H_1 (H_2) be obtained by deleting the first row of a $b \times b$ $(v \times v)$ orthogonal matrix for which all the elements in the first row are equal. Then

$$H_1H_1' = I_{b-1}, \quad H_2H_2' = I_{v-1}.$$

Let W be the matrix of coefficients for the equations to estimate the contrasts of block effects vector β and treatment effects vector τ after eliminating μ, given by

$$W = \begin{pmatrix} kH_1H_1' & H_1N'H_2' \\ H_2NH_1' & rH_2H_2' \end{pmatrix} = \begin{pmatrix} kI_{b-1} & H_1N'H_2' \\ H_2NH_1' & rI_{v-1} \end{pmatrix}. \tag{7.7}$$

Now

$$|W| = r^{v-1}\left| kI_{b-1} - \frac{1}{r}H_1N'H_2'H_2NH_1' \right|. \tag{7.8}$$

For fixed v, b, r, k, if $\left| kI_{b-1} - \frac{1}{r}H_1N'H_2'H_2NH_1' \right|$ is maximized, then $|W|$ will be maximized and consequently $\left| rI_{v-1} - \frac{1}{k}H_2NH_1'H_1N'H_2' \right|$ will be maximized. The matrices of coefficients for estimating contrasts of block and treatment effects are respectively

$$M_1 = kI_{b-1} - \frac{1}{r}H_1N'H_2'H_2NH_1', \text{ and}$$

$$M_2 = rI_{v-1} - \frac{1}{k}H_2NH_1'H_1N'H_2'.$$

Furthermore, from Sec. 2.8, we know that LB is D-optimal for the estimation of block effects. Hence the D-optimality for the estimation of treatment effects.

Again

$$W^{-1} = \begin{pmatrix} M_1^{-1} & -M_1^{-1}H_1N'H_2' \\ -H_2NH_1'M_1^{-1} & \frac{1}{r}I_{v-1} + \frac{1}{r^2}H_2NH_1'M_1^{-1}H_1N'H_2' \end{pmatrix} \tag{7.9}$$

and it can be verified that

$$\text{Trace } W^{-1} = \{\text{Trace } M_1^{-1}\}\left(1 + \frac{k}{r}\right) + \left(\frac{v-b}{r}\right)$$

$$= \{\text{Trace } M_2^{-1}\}\left(1 + \frac{r}{k}\right) + \left(\frac{b-v}{k}\right). \tag{7.10}$$

For a LB design, the block effects are A-optimally estimated (Sec. 2.8) and hence trace M_1^{-1} is minimum, which implies that trace W^{-1} is minimum and trace M_2^{-1} is minimum. Thus the treatment effects are A-optimally estimated for a LB design.

E-optimality can be similarly demonstrated by showing that the smallest eigenvalues of M_1, W and M_2 are the same.

7.4 Application of LB Designs in Successive Sampling

In successive sampling, sampling will be done on two occasions keeping a portion of the sample common for both occasions. Let N be the population size, n be the sample sizes on both occasions and $m(<n)$ be the common sample size for both occasions. We are interested in developing an estimate of the population mean at the second occasion as a linear combination of two estimators: one based on the matched sample and the other based on the unmatched sample. The responses collected on the first occasion will be considered as auxiliary information to construct estimators based on regression, ratio, or difference methods for the matched sample. For further details on successive sampling, we refer to Sukhatme and Sukhatme (1970).

Singh and Raghavarao (1975) suggested the use of LB designs in successive sampling. Suppose a LB design exists with parameters $v = N, b, r, k = n, \mu = m$. The population units will be identified with the symbols of the design. A set of the LB design will be selected with probability $1/b$ as the first sample, and from the remaining $b - 1$ sets, one set will be selected with probability $1/(b - 1)$ as the second sample. Let X and Y denote the response variables for the first and second occasions, respectively.

Let $\bar{x}_n, \bar{y}_n$ be the sample means on two occasions; $\bar{x}_m, \bar{y}_m$ be the sample means for the matched portion; and $\bar{y}_{n-m}$ be the sample mean for the unmatched portion. Let β be the population regression coefficient of Y on X, estimated by $\hat{\beta}$ from the matched sample. Finally, let $\bar{X}$ and $\bar{Y}$ be the population means.

Let $\hat{\bar{Y}}_1$ be the estimator for the matched sample, given by

$$\hat{\bar{Y}}_1 = \bar{y}_m + \hat{\beta}(\bar{x}_n - \bar{x}_m) \tag{7.11}$$

based on regression method, and let $\hat{\bar{Y}}_2 = \bar{y}_{n-m}$, be the ordinary sample mean based on the unmatched sample. Combining the two estimators, we get the LB estimator

$$\hat{\bar{Y}}_{\mathrm{LB}} = w_1 \hat{\bar{Y}}_1 + w_2 \hat{\bar{Y}}_2, \tag{7.12}$$

where

$$w_i = \frac{1/\mathrm{Var}(\hat{\bar{Y}}_i)}{\sum_{j=1}^{2} 1/\mathrm{Var}(\hat{\bar{Y}}_j)}, \quad i = 1, 2. \tag{7.13}$$

Singh and Raghavarao (1975) showed the following results:

Theorem 7.3

$$E\left(\hat{\bar{Y}}_{\text{LB}}\right) = \bar{Y} - w_1 E_1 \text{Cov}_2\left(\hat{\beta}, \bar{x}_m\right),\tag{7.14}$$

the subscripts 1 and 2 denoting the expectation for the first occasion and covariance for the second occasion.

From (7.14), the estimator $\hat{\bar{Y}}_{\text{LB}}$ is a biased estimator of $\bar{Y}$. However, the bias is zero if the joint distribution of X and Y is nearly bivariate normal (Sukhatme and Sukhatme, 1970, p. 195).

Let λ_{ij} be the number of sets of the LB in which symbols i and j occur together, and let

$$V(\bar{y}_m) = \frac{1}{Nm}\sum_{i=1}^{N} y_i^2 + \frac{1}{m^2 b(b-1)}\sum_{\substack{i,j=1\\i\neq j}}^{N} y_i y_j \lambda_{ij}(\lambda_{ij}-1) - \bar{Y}^2,$$

$$\tag{7.15}$$

$$V(\bar{y}_{n-m}) = \frac{1}{N(n-m)}\sum_{i=1}^{N} y_i^2$$

$$+ \frac{1}{(n-m)^2 b(b-1)}\sum_{\substack{i,j=1\\i\neq j}}^{N} y_i y_j \lambda_{ij}(b-2r+\lambda_{ij}) - \bar{Y}^2,$$

$$\tag{7.16}$$

$$V(\bar{x}_m) = \frac{1}{Nm}\sum_{i=1}^{N} x_i^2 + \frac{1}{m^2 b(b-1)}\sum_{\substack{i,j=1\\i\neq j}}^{N} x_i x_j \lambda_{ij}(\lambda_{ij}-1) - \bar{X}^2,$$

$$\tag{7.17}$$

$$\text{Cov}(\bar{x}_n, \bar{x}_m) = V(\bar{x}_n) = \frac{1}{Nn}\sum_{i=1}^{N} x_i^2 + \frac{1}{n^2 b}\sum_{\substack{i,j=1\\i\neq j}}^{N} x_i x_j \lambda_{ij} - \bar{X}^2,\tag{7.18}$$

$$\text{Cov}(\bar{x}_n, \bar{y}_m) = \text{Cov}(\bar{x}_n, \bar{y}_n) = \frac{1}{Nn}\sum_{i=1}^{N} x_i y_i + \frac{1}{n^2 b}\sum_{\substack{i,j=1\\i\neq j}}^{N} x_i y_j \lambda_{ij} - \bar{X}\bar{Y},$$

$$\tag{7.19}$$

$$\mathrm{Cov}(\bar{x}_m, \bar{y}_m) = \frac{1}{Nm} \sum_{i=1}^{N} x_i y_i + \frac{1}{m^2 b(b-1)} \sum_{\substack{i,j=1 \\ i \neq j}}^{N} x_i y_j \lambda_{ij}(\lambda_{ij} - 1) - \bar{X}\bar{Y}.$$

$$(7.20)$$

Then

Theorem 7.4

$$V\left(\hat{\bar{Y}}_1\right) = V(\bar{y}_m) + \beta^2\{V(\bar{x}_m) - V(\bar{x}_n)\} - 2\beta\{\mathrm{Cov}(\bar{x}_m, \bar{y}_m)$$
$$-\mathrm{Cov}(\bar{x}_n, \bar{y}_n)\},$$
$$V\left(\hat{\bar{Y}}_2\right) = V(\bar{y}_{n-m}), \qquad (7.21)$$

where the expressions on the right are given before the theorem.

The variance of $\hat{\bar{Y}}_{\mathrm{LB}}$ along with its estimator is given by Singh and Raghavarao (1975).

8

Partially Balanced Incomplete Block Designs

8.1 Definitions and Preliminaries

We noted in Chap. 4 that variance balanced block designs have many useful properties among block designs. However, they do not exist for all parametric combinations. By relaxing variance balancedness and allowing different variances for estimated elementary contrasts of treatment effects, Bose and Nair (1939) introduced Partially Balanced Incomplete Block (PBIB) designs. To make a mathematically tractable statistical analysis for these designs and study them systematically, we need the concept of association scheme on v symbols as defined below:

Definition 8.1 Given v symbols $1, 2, \ldots, v$, a relation satisfying the following conditions is said to be an *association scheme with m classes*:

1. Any two symbols α and β are either first, second, $\ldots$, or mth associates and this relationship is symmetrical. We denote $(\alpha, \beta) = i$, when α and β are ith associates.
2. Each symbol α has n_i, ith associates, the number n_i being independent of α.
3. If $(\alpha, \beta) = i$, the number of symbols γ that satisfy simultaneously $(\alpha, \gamma) = j$, $(\beta, \gamma) = j'$ is $p^i_{jj'}$ and this number is independent of α and β. Further, $p^i_{jj'} = p^i_{j'j}$.

The numbers $v, n_i, p^i_{jj'}$ are called the parameters of the association scheme. The parameters $p^i_{jj'}$ can be written in m matrices of order $m \times m$ as follows:

$$P_i = (P^i_{jj'}), \quad i = 1, 2, \ldots, m; \quad j, j' = 1, 2, \ldots, m. \tag{8.1}$$

Given an m-class association scheme on v symbols, a PBIB design with m associate classes is defined in Definition 8.2.

Definition 8.2 A PBIB design with m associate classes is an arrangement of v symbols in b sets of size $k (< v)$ such that

1. Every symbol occurs at most once in a set.
2. Every symbol occurs in r sets.

3. Two symbols α and β, occur in λ_i sets, if $(\alpha, \beta) = i$ and λ_i is independent of the symbols α and β.

The numbers v, b, r, k, λ_i are the parameters of the PBIB design. The PBIB design is usually identified by the association scheme of the symbols. For example, a group divisible (GD) design is a PBIB design where the symbols have a group divisible association scheme. The parameters of the association scheme and design satisfy the following relations:

$$vr = bk, \quad \sum_{i=1}^{m} n_i = v - 1, \quad \sum_{i=1}^{n} n_i \lambda_i = r(k - 1),$$

$$\sum_{j'=1}^{m} p_{jj'}^i = n_j - \delta_{jj'}, \quad n_i p_{jj'}^i = n_j p_{ij'}^j = n_{j'} p_{ij}^{j'}, \tag{8.2}$$

where $\delta_{jj'} = 1(0)$, according to $j = j'$ $(j \neq j')$. The proofs of the relations (8.2) can be found in Raghavarao (1971).

Let us define m matrices $B_i = (b_{\alpha\beta}^i)$, $i = 1, 2, \ldots, m$ of order $v \times v$, where

$$b_{\alpha\beta}^i = \begin{cases} 1, & \text{if } (\alpha, \beta) = i \\ 0, & \text{if } (\alpha, \beta) \neq i. \end{cases} \tag{8.3}$$

In addition, let $B_0 = I_v$. The matrices $B_0, B_1, \ldots, B_m$ are called the association matrices, of the association scheme, introduced by Bose and Mesner (1959) and were shown by them to be independent and commutative satisfying the relations

$$\sum_{i=0}^{m} B_i = J_v; \quad B_j B_{j'} = \sum_{i=0}^{m} p_{jj'}^i B_i; \quad j, j' = 0, 1, \ldots, m,$$

$$\text{where } p_{jj'}^0 = n_j \delta_{jj'}. \tag{8.4}$$

From Definition 8.2, we have

$$NN' = r B_0 + \lambda_1 B_1 + \cdots + \lambda_m B_m. \tag{8.5}$$

Theorem 8.1, first proved by Connor and Clatworthy (1954) and later elegantly proved by Bose and Meser (1959), will be useful in determining the eigenvalues of NN' (also see Raghavarao, 1971).

Theorem 8.1 *The distinct eigenvalues of NN' are the same as the distinct eigenvalues of*

$$P^* = r I_{m+1} + \lambda_1 P_1^* + \cdots + \lambda_m P_m^*, \tag{8.6}$$

where $p_{0j'}^i = \delta_{ij'}$ and

$$P_i^* = \begin{pmatrix} 0 & p_{01}^i & p_{02}^i & \cdots & p_{0m}^i \\ p_{10}^i & & & & \\ p_{20}^i & & P_i & & \\ \vdots & & & & \\ p_{m0}^i & & & & \end{pmatrix}.$$

Let us now determine the eigenvalues along with their multiplicities of NN' for a connected PBIB design with two associate classes. The eigenvalues of P^* can be easily determined to be

$$\theta_0 = rk, \quad \theta_i = r - 1/2\{(\lambda_1 - \lambda_2)[-u + (-1)^i \sqrt{\Delta}] + (\lambda_1 + \lambda_2)\}, \quad i = 1, 2, \tag{8.7}$$

where

$$u = p_{12}^2 - p_{12}^1, \quad w = p_{12}^2 + p_{12}^1, \quad \Delta = u^2 + 2w + 1. \tag{8.8}$$

Then θ_0, θ_1 and θ_2 are the distinct eigenvalues of NN'. Let α_0, α_1 and α_2 be the respective multiplicities. Since the design is connected, $\alpha_0 = 1$. Furthermore,

$$\alpha_1 + \alpha_2 = v - 1,$$

$$\text{tr}(NN') = vr = rk + \alpha_1\theta_1 + \alpha_2\theta_2,$$

where tr is the trace of the matrix. Solving the above equations we get

$$\alpha_i = \frac{n_1 + n_2}{2} + (-1)^i \left[\frac{(n_1 - n_2) + u(n_1 + n_2)}{2\sqrt{\Delta}} \right], \quad i = 1, 2. \tag{8.9}$$

It is to be noted that the multiplicities depend only on the parameters of the association scheme, not on the parameters of the design.

While the parameters are uniquely determined from the association scheme, the association scheme may or may not be uniquely determined from the parameters. For some results on the uniqueness of association schemes, we refer to Raghavarao (1971).

In the next section, we will discuss some known two and higher class association schemes and in Sec. 8.3 we will give the intra-block analysis of two and three associate class PBIB designs. In the subsequent sections, we will examine the combinatorics and applications of some of these designs.

8.2 Some Known Association Schemes

We will define some useful known association schemes, give their parameters and eigenvalues and their multiplicities of NN' of the corresponding PBIB designs.

Definition 8.3 In a group divisible association scheme there are $v = mn$ symbols arranged in m groups of n symbols. Two symbols in the same group are first associates and two symbols in different groups are second associates.

Clearly

$$n_1 = n - 1, \quad n_2 = n(m - 1),$$

$$P_1 = \begin{pmatrix} n - 2 & 0 \\ 0 & n(m - 1) \end{pmatrix}, \quad P_2 = \begin{pmatrix} 0 & n - 1 \\ n - 1 & n(m - 2) \end{pmatrix}. \tag{8.10}$$

The three distinct eigenvalues of NN' are $\theta_0 = rk$, $\theta_1 = r - \lambda_1$, and $\theta_2 = rk - v\lambda_2$ with the respective multiplicities $\alpha_0 = 1$, $\alpha_1 = m(n - 1)$, $\alpha_2 = m - 1$.

Definition 8.4 In a triangular association scheme, there are $v = n(n - 1)/2$ symbols arranged in an $n \times n$ array above the diagonal, leaving the diagonal blank and symmetrically filling the symbols below the diagonal. Two symbols occurring in the same row or column are first associates and two symbols not occurring in the same row or column are second associates.

Clearly

$$n_1 = 2(n - 2), \quad n_2 = (n - 2)(n - 3)/2,$$

$$P_1 = \begin{pmatrix} n - 2 & n - 3 \\ n - 3 & (n - 3)(n - 4)/2 \end{pmatrix}, \quad P_2 = \begin{pmatrix} 4 & 2n - 8 \\ 2n - 8 & (n - 4)(n - 5)/2 \end{pmatrix}. \tag{8.11}$$

The three distinct eigenvalues of NN' are $\theta_0 = rk$, $\theta_1 = r + (n - 4)\lambda_1 - (n - 3)\lambda_2$, and $\theta_2 = r - 2\lambda_1 + \lambda_2$ with the respective multiplicities $\alpha_0 = 1$, $\alpha_1 = n - 1$, and $\alpha_2 = n(n - 3)/2$.

Definition 8.5 In an L_i association scheme, there are $v = s^2$ symbols arranged in an $s \times s$ square array and $i - 2$ mutually orthogonal Latin squares are superimposed on the square array. Two symbols are first associates if they occur in the same row or column of the array or in positions occupied by the same letter in any of the $i - 2$ Latin squares, and other pairs of symbols are second associates.

Clearly

$$n_1 = i(s-1), \quad n_2 = (s-i+1)(s-1),$$

$$P_1 = \begin{pmatrix} (i-1)(i-2)+s-2 & (s-i+1)(i-1) \\ (s-i+1)(i-1) & (s-i+1)(s-i) \end{pmatrix}, \tag{8.12}$$

$$P_2 = \begin{pmatrix} i(i-1) & i(s-i) \\ i(s-i) & (s-i)(s-i-1)+s-2 \end{pmatrix}.$$

The three distinct eigenvalues of NN' are $\theta_0 = rk$, $\theta_1 = r + (s-i)\lambda_1 - (s-i+1)\lambda_2$, and $\theta_2 = r - i\lambda_1 + (i-1)\lambda_2$ with the respective multiplicities $\alpha_0 = 1$, $\alpha_1 = i(s-1)$, and $\alpha_2 = (s-i+1)(s-1)$.

Definition 8.6 In a cyclic association scheme, the first associates of ith symbol are $(i + d_1, i + d_2, \ldots, i + d_{n_1})$ mod v, while all other symbols are second associates, where the n_1, d-elements satisfy

1. The d_j are all distinct and $0 < d_j < v$, for $j = 1, 2, \ldots, n_1$.
2. Among the $n_1(n_1 - 1)$ differences $d_j - d_{j'} \pmod{v}$, each of the $d_1, d_2, \ldots, d_{n_1}$ elements occurs p_{11}^1 times and each of the $e_1, e_2, \ldots, e_{n_2}$ elements occur p_{11}^2 times where $d_1, d_2, \ldots, d_{n_1}, e_1, e_2, \ldots, e_{n_2}$ are distinct nonzero elements of the module M of v elements $0, 1, \ldots, v - 1$.
3. The set $D = \{d_1, d_2, \ldots, d_{n_1}\} = \{-d_1, -d_2, \ldots, -d_{n_1}\}$.

All known cyclic association schemes have the parameters

$$v = 4t + 1, \quad n_1 = n_2 = 2t,$$

$$P_1 = \begin{pmatrix} t-1 & t \\ t & t \end{pmatrix}, \quad P_2 = \begin{pmatrix} t & t \\ t & t-1 \end{pmatrix}. \tag{8.13}$$

The two associate class PBIB designs were classified by Bose and Shimamoto (1952) into the following types depending on the association scheme:

1. Group Divisible (GD);
2. Simple (S);
3. Triangular (T);
4. Latin Square Type (L_i); and
5. Cyclic.

We defined association schemes for $1, 3, 4$ and 5 types of designs. Simple PBIB designs have either $\lambda_1 = 0, \lambda_2 \neq 0$, or $\lambda_1 \neq 0, \lambda_2 = 0$. The known simple designs are either partial geometric designs as defined by Bose (1963) or replications of partial geometric designs.

The two-associate-class association schemes that are not covered by Bose and Shimamoto classification are of two categories. The first category consists of pseudo-triangular, pseudo-Latin-square type and pseudo-cyclic having the parameters given by (8.11), (8.12) and (8.13), with any combinatorial structure. The other category are neither simple, nor their parameters satisfy any previously listed schemes. The NL_j family of designs with parameters given in (8.12), where s and i are replaced by $-t$ and $-j$ was studied by Mesner (1967). Other examples for $v = 15$ was given by Clatworthy (1973) and for $v = 50$ was given by Hoffman and Singleton (1960).

Bose and Connor (1952) subdivided the GD designs into

(i) Singular GD designs with $r - \lambda_1 = 0, rk - v\lambda_2 > 0$.
(ii) Semi-regular GD designs with $r - \lambda_1 > 0, rk - v\lambda_2 = 0$.
(iii) Regular GD designs with $r - \lambda_1 > 0, rk - v\lambda_2 > 0$.

These three classes of GD designs possess different combinatorial properties and we will discuss some of them in Sec. 8.4.

The updated tables of two-associate-class PBIB designs prepared by Clatworthy (1973) lists the parameters and plans for 124 singular GD designs, 110 semi-regular GD designs, 200 regular GD designs, 100 triangular designs, 146 Latin-square-type designs, 29 cyclic-type designs, 15 partial geometric designs, and 42 miscellaneous designs, which do not fit into any of the previous categories.

In Illustrations 8.1, 8.2 and 8.3 we will give examples of GD, triangular, and cyclic designs.

Illustration 8.1 Let $v = 6, m = 3, n = 2$ and the 6 symbols 0, 1, 2, 3, 4, 5 are arranged in 3 groups of 2 symbols each as $\{0, 1\}, \{2, 3\}, \{4, 5\}$.

$$\text{Here } (0, 1) = (2, 3) = (4, 5) = 1;$$
$$(0, 2) = (0, 3) = (0, 4) = (0, 5) = (1, 2) = (1, 3) = (1, 4)$$
$$= (1, 5) = (2, 4) = (2, 5) = (3, 4) = (3, 5) = 2.$$

The arrangement of 6 symbols in 3 sets

$$(0, 1, 2, 3)$$
$$(0, 1, 4, 5)$$
$$(2, 3, 4, 5)$$

is a GD design with parameters $v = 6, m = 3, n = 2, b = 3, r = 2, k = 4,$ $\lambda_1 = 2, \lambda_2 = 1$.

Illustration 8.2　Let the $v = 6$ symbols be arranged in a 4×4 array as follows:

$$
\begin{array}{cccc}
- & 0 & 1 & 2 \\
0 & - & 3 & 4 \\
1 & 3 & - & 5 \\
2 & 4 & 5 & -
\end{array}
$$

Here $(0, 1) = (0, 2) = (0, 3) = (0, 4) = (1, 2) = (1, 3) = (1, 5) = (2, 4) = (2, 5) = (3, 4) = (3, 5) = (4, 5) = 1$; $(0, 5) = (1, 4) = (2, 3) = 2$.

The arrangement of 6 symbols in 3 sets

$$
\begin{aligned}
& (0, 1, 4, 5) \\
& (0, 2, 3, 5) \\
& (1, 2, 3, 4)
\end{aligned}
$$

is a triangular design with parameters $v = 6, n = 4, b = 3, r = 2, k = 4, \lambda_1 = 1, \lambda_2 = 2$.

Illustration 8.3　Let $v = 5$, with $t = 1$ and $M = \{0, 1, 2, 3, 4\}$. Let $D = \{1, 4\}$, so that $(0, 1) = (0, 4) = (1, 2) = (2, 3) = (3, 4) = 1$; $(0, 2) = (0, 3) = (1, 3) = (1, 4) = (2, 4) = 2$.

The arrangement of 5 symbols in 5 sets

$$
\begin{aligned}
& (0, 1, 2) \\
& (1, 2, 3) \\
& (2, 3, 4) \\
& (3, 4, 0) \\
& (4, 0, 1)
\end{aligned}
$$

is a cyclic design with parameters $v = 5, t = 1, b = 5, r = 3, k = 3, \lambda_1 = 2, \lambda_2 = 1$.

We will now give some association schemes with more than two classes.

Definition 8.7　(Vartak, 1955) In a rectangular association scheme there are $v = mn$ symbols arranged in an $m \times n$ rectangle. Two symbols occurring in the same row are first associates; occurring in the same column are second associates; not occurring in the same row or column are third associates.

Clearly

$$n_1 = n - 1, \quad n_2 = m - 1, \quad n_3 = (n-1)(m-1),$$

$$P_1 = \begin{pmatrix} n-2 & 0 & 0 \\ 0 & 0 & m-1 \\ 0 & m-1 & (m-1)(n-2) \end{pmatrix},$$

$$P_2 = \begin{pmatrix} 0 & 0 & n-1 \\ 0 & m-2 & 0 \\ n-1 & 0 & (m-2)(n-1) \end{pmatrix}, \qquad (8.14)$$

$$P_3 = \begin{pmatrix} 0 & 1 & n-2 \\ 1 & 0 & m-2 \\ n-2 & m-2 & (m-2)(n-2) \end{pmatrix}.$$

The four distinct eigenvalues of NN' are $\theta_0 = rk$, $\theta_1 = (r - \lambda_1) + (m-1)(\lambda_2 - \lambda_3)$, $\theta_2 = r - \lambda_2 + (n-1)(\lambda_1 - \lambda_3)$ and $\theta_3 = r - \lambda_1 - \lambda_2 + \lambda_3$ with the respective multiplicities $\alpha_0 = 1$, $\alpha_1 = n - 1$, $\alpha_2 = m - 1$, $\alpha_3 = (m-1)(n-1)$.

Definition 8.8 (Raghavarao and Chandrasekhararao, 1964) In the cubic association scheme there are $v = s^3$ symbols denoted by (α, β, γ) for $\alpha, \beta, \gamma = 1, 2, \ldots, s$. The distance δ between two symbols (α, β, γ) and $(\alpha', \beta', \gamma')$ is the number of nonzero elements in $(\alpha - \alpha', \beta - \beta', \gamma - \gamma')$. Two symbols (α, β, γ) and $(\alpha', \beta', \gamma')$ are ith associates, if the distance between them is $\delta = i$, for $i = 1, 2, 3$.

Clearly

$$n_1 = 3(s-1), \quad n_2 = 3(s-1)^2, \quad n_3 = (s-1)^3,$$

$$P_1 = \begin{pmatrix} s-2 & 2(s-1) & 0 \\ 2(s-1) & 2(s-1)(s-2) & (s-1)^2 \\ 0 & (s-1)^2 & (s-1)^2(s-2) \end{pmatrix},$$

$$P_2 = \begin{pmatrix} 2 & 2(s-2) & s-1 \\ 2(s-2) & 2(s-1)+(s-2)^2 & 2(s-1)(s-2) \\ s-1 & 2(s-1)(s-2) & (s-1)(s-2)^2 \end{pmatrix}, \qquad (8.15)$$

$$P_3 = \begin{pmatrix} 0 & 3 & 3(s-2) \\ 3 & 6(s-2) & 3(s-2)^2 \\ 3(s-2) & 3(s-2)^2 & (s-2)^3 \end{pmatrix}.$$

The four distinct eigenvalues of NN' are $\theta_0 = rk$,

$$\theta_1 = r + (2s-3)\lambda_1 + (s-1)(s-3)\lambda_2 - (s-1)^2\lambda_3,$$

$$\theta_2 = r + (s-3)\lambda_1 - (2s-3)\lambda_2 + (s-1)\lambda_3, \text{ and}$$

$$\theta_3 = r - 3\lambda_1 + 3\lambda_2 - \lambda_3$$

with the respective multiplicities

$$\alpha_0 = 1, \quad \alpha_1 = 3(s-1), \quad \alpha_2 = 3(s-1)^2, \quad \alpha_3 = (s-1)^3.$$

Definition 8.9 (P.W.M. John, 1966; Bose and Laskar, 1967) In the *Extended Triangular Association Scheme* there are $v = (s+2)(s+3)(s+4)/6$ symbols represented by (α, β, γ) for $0 < \alpha < \beta < \gamma \le s+4$. Two symbols are first associates if they have two integers in common, second associates if they have one integer in common, and third associates otherwise.

Clearly

$$n_1 = 3(s+1), \quad n_2 = 3s(s+1)/2, \quad n_3 = s(s-1)(s+1)/6,$$

$$P_1 = \begin{pmatrix} s+2 & 2s & 0 \\ 2s & s^2 & s(s-1)/2 \\ 0 & s(s-1)/2 & s(s-1)(s-2)/6 \end{pmatrix},$$

$$P_2 = \begin{pmatrix} 4 & 2s & s-1 \\ 2s & (s-1)(s+6)/2 & (s-1)(s-2) \\ s-1 & (s-1)(s-2) & (s-1)(s-2)(s-3)/6 \end{pmatrix}, \quad (8.16)$$

$$P_3 = \begin{pmatrix} 0 & 9 & 3(s-2) \\ 9 & 9(s-2) & 3(s-2)(s-3)/2 \\ 3(s-2) & 3(s-2)(s-3)/2 & (s-2)(s-3)(s-4)/6 \end{pmatrix}.$$

The four distinct eigenvalues of NN' are $\theta_0 = rk$,

$$\theta_1 = r + (2s-1)\lambda_1 + [s(s-5)/2]\lambda_2 - [s(s-1)/2]\lambda_3,$$
$$\theta_2 = r + (s-3)\lambda_1 - (2s-3)\lambda_2 + (s-1)\lambda_3, \quad \text{and}$$
$$\theta_3 = r - 3\lambda_1 + 3\lambda_2 - \lambda_3$$

with the respective multiplicities

$$\alpha_0 = 1, \quad \alpha_1 = s+3, \quad \alpha_2 = (s+1)(s+4)/2, \quad \alpha_3 = (s-1)(s+3)(s+4)/6.$$

Definition 8.10 (Singla, 1977a) In *Extended L_2 Association Scheme*, there are $v = ms^2$ symbols represented by (i, α, β), for $i = 1, 2, \ldots, m; \alpha, \beta = 1, 2, \ldots, s$. For the symbol (i, α, β) the first associates are (i', α, β) for $i' = 1, 2, \ldots, m$; $i' \ne i$; the second associates are $(i', \alpha', \beta), (i', \alpha, \beta')$ for $i' = 1, 2, \ldots, m; \alpha', \beta' = 1, 2, \ldots, s; \alpha' \ne \alpha, \beta' \ne \beta$, and all other symbols are third associates.

Clearly

$$n_1 = m - 1, \quad n_2 = 2m(s - 1), \quad n_3 = m(s - 1)^2,$$

$$P_1 = \begin{pmatrix} m - 2 & 0 & 0 \\ 0 & 2m(s - 1) & 0 \\ 0 & 0 & m(s - 1)^2 \end{pmatrix},$$

$$P_2 = \begin{pmatrix} 0 & m - 1 & 0 \\ m - 1 & m(s - 2) & m(s - 1) \\ 0 & m(s - 1) & m(s - 1)(s - 2) \end{pmatrix}, \qquad (8.17)$$

$$P_3 = \begin{pmatrix} 0 & 0 & m - 1 \\ 0 & 2m & 2m(s - 2) \\ m - 1 & 2m(s - 2) & m(s - 2)^2 \end{pmatrix}.$$

The four distinct eigenvalues of NN' are $\theta_0 = rk$, $\theta_1 = r - \lambda_1$, $\theta_2 = r + (m - 1)\lambda_1 + m(s - 2)\lambda_2 - m(s - 1)\lambda_3$, and $\theta_3 = r + (m - 1)\lambda_1 - 2m\lambda_2 + m\lambda_3$, with the respective multiplicities $\alpha_0 = 1$, $\alpha_1 = s^2(m - 1)$, $\alpha_2 = 2(s - 1)$, $\alpha_3 = (s - 1)^2$.

Definition 8.11 (Tharthare, 1965) In a *Generalized Right Angular Association Scheme*, there are $v = p\ell s$ symbols represented by (α, β, γ), where $\alpha = 1, 2, \ldots, \ell$; $\beta = 1, 2, \ldots, p$, $\gamma = 1, 2, \ldots, s$. For the symbol (α, β, γ) the first associates are (α, β, γ') for $\gamma \neq \gamma'$, the second associates are $(\alpha, \beta', \gamma')$ for $\beta' = 1, 2 \ldots, p$; $\gamma' = 1, 2, \ldots, s$; $\beta' \neq \beta$; third associates are $(\alpha', \beta, \gamma')$ for $\alpha' = 1, 2, \ldots, \ell$; $\gamma' = 1, 2, \ldots, s$; $\alpha' \neq \alpha$; and all other symbols are fourth associates.

Clearly

$$n_1 = s - 1, \quad n_2 = s(p - 1), \quad n_3 = s(\ell - 1), \quad n_4 = s(\ell - 1)(p - 1),$$

$$P_1 = \begin{pmatrix} s - 2 & 0 & 0 & 0 \\ 0 & s(p - 1) & 0 & 0 \\ 0 & 0 & s(\ell - 1) & 0 \\ 0 & 0 & 0 & s(\ell - 1)(p - 1) \end{pmatrix},$$

$$P_2 = \begin{pmatrix} 0 & s - 1 & 0 & 0 \\ s - 1 & s(p - 2) & 0 & 0 \\ 0 & 0 & 0 & s(\ell - 1) \\ 0 & 0 & s(\ell - 1) & s(\ell - 1)(p - 2) \end{pmatrix}, \qquad (8.18)$$

$$P_3 = \begin{pmatrix} 0 & 0 & s-1 & 0 \\ 0 & 0 & 0 & s(p-1) \\ s-1 & 0 & s(\ell-2) & 0 \\ 0 & s(p-1) & 0 & s(p-1)(\ell-2) \end{pmatrix},$$

$$P_4 = \begin{pmatrix} 0 & 0 & 0 & s-1 \\ 0 & 0 & s & s(p-2) \\ 0 & s & 0 & s(\ell-2) \\ s-1 & s(p-2) & s(\ell-2) & s(\ell-2)(p-2) \end{pmatrix}.$$

The five distinct eigenvalues of NN' are

$$\theta_0 = rk, \quad \theta_1 = r - \lambda_1 + s(\lambda_1 - \lambda_2) + s(\ell - 1)(\lambda_3 - \lambda_4),$$
$$\theta_2 = r - \lambda_1, \quad \theta_3 = r - \lambda_1 + s[(\lambda_1 - \lambda_3) + (p - 1)(\lambda_2 - \lambda_4)], \quad \text{and}$$
$$\theta_4 = r - \lambda_1 + s(\lambda_1 - \lambda_2 - \lambda_3 + \lambda_4),$$

with the respective multiplicities $\alpha_0 = 1, \alpha_1 = p - 1, \alpha_2 = p\ell(s - 1), \alpha_3 = \ell - 1,$ $\alpha_4 = (p - 1)(\ell - 1)$.

A particular case of this scheme when $p = 2$ is the Right Angular Association Scheme, also due to Tharthare (1963).

Raghavarao and Aggarwal (1973) defined the extended right angular association scheme in 7 associate classes with $v = n_1 n_2 n_3 n_4$ symbols.

A polygonal association scheme is defined by Frank and O'Shaughnessy (1974). Let there be $v = sn$ symbols represented by (α, β) where $\alpha = 0, 1, \ldots, s - 1$ and $\beta = 0, 1, \ldots, n - 1$. Two symbols (α, β) and (α', β') are ith associates if and only if $\alpha \equiv \alpha' + i - 1 \pmod{s}$ or $\alpha' \equiv \alpha + i - 1 \pmod{s}$. This relationship may be interpreted geometrically by thinking that the v symbols lie at the vertices of n nested s-sided regular polygons such that n vertices are collinear. Consider lines each of which joins such a set of n vertices. Two symbols are first associates if they lie on the same line, second associates if they lie on adjacent lines, third associates if they are separated by exactly one line, fourth associates if they are separated by exactly 2 lines, and so on. When s is odd we get $m = (s + 1)/2$ associate classes and when s is even we get $m = s/2 + 1$ associate classes. The design is called a triangle, square, pentagon, or hexagon design depending on $s = 3, 4, 5,$ or $6,$ respectively.

We will consider Residue Classes Association Scheme. Let $v = tm + 1$ be a prime or prime power. The element $\alpha \in GF(v)$ is said to be the tth order residue of $GF(v)$ if the congruence equation

$$x^t \equiv \alpha \pmod{v} \tag{8.19}$$

has a solution. Let x be a primitive root of GF(v). Let $H = \{x^0, x^t, \ldots, x^{(m-1)t}\}$ be the set of tth order residues. Two elements $\alpha, \beta \in$ GF(v) may be called ith associates if $\alpha - \beta \in x^{i-1} H$ and the association relation is symmetric if $-1 \in H$. Thus when $-1 \in H$, we can define an association scheme called residue classes association scheme. This scheme was introduced and studied by Aggarwal and Raghavarao (1972).

Let $v = n_1 n_2 \cdots n_m$ and let the v symbols be denoted by $(\alpha_1, \alpha_2, \ldots, \alpha_m)$, where $\alpha_i = 0, 1, \ldots, n_{i-1}$ for $i = 1, 2, \ldots, m$. Raghavarao (1960b) defined generalized group divisible association scheme, where the ith associates of any symbol are those symbols having the same first $(m - i)$ elements. Depending on the number and positions of identical elements in two symbols, Hinkelmann and Kempthorne (1963) defined Extended Group Divisible Association Scheme with $2^m - 1$ associate classes and these designs are useful as balanced confounded factorial experiments (see Shah, 1958).

The cubic association scheme has been generalized by Kusumoto (1965) and the triangular association scheme has been generalized by Ogasawara (1965). For other higher association schemes, the interested reader is referred to Adhikary (1966, 1967), and Yamamoto, Fuji, and Hamada (1965). Kusumoto (1967) and Surendran (1968) consider Kronecker Product Association Scheme by taking the Kronecker product of association matrices of two different association schemes.

Kageyama (1972c, 1974b, c, d) and Vartak (1955) discussed the problem of reducing more associate class schemes into fewer associate class schemes.

8.3 Intra-Block Analysis

Assuming the model and notation used in Chap. 2, we have the reduced normal equations estimating $\hat{\tau}$ given by

$$C_{\tau|\beta} \hat{\tau} = \mathbf{Q}_{\tau|\beta}, \tag{8.20}$$

where

$$\mathbf{Q}_{\tau|\beta} = \mathbf{T} - \frac{1}{k} N \mathbf{B},$$

$$C_{\tau|\beta} = r B_0 - (1/k) \left\{ r B_0 + \sum_{i=1}^{m} \lambda_i B_i \right\},$$

$$= \frac{a}{k} B_0 - \frac{1}{k} \sum_{i=1}^{m} \lambda_i B_i,$$

and $a = r(k - 1)$.

Let Q_i be the adjusted treatment total of the ith treatment, which is the ith component of the v-dimensional vector $\mathbf{Q}_{\tau|\beta}$. Let $S_j(Q_i)$ and $S_j(\hat{\tau}_i)$ denote the sum of Q_i's and $\hat{\tau}_i$'s respectively over the n_j, jth associates of the ith treatment. The following lemma, whose proof is obvious will be required to solve Eq. (8.20).

Lemma 8.1 *If θ is any treatment and G_i is the set of the ith associates of θ and if $G_{i,j}$ is the collection of treatments which are jth associates of each treatment of G_i; then*

(i) *θ occurs $n_i \delta_{ij}$ times in $G_{i,j}$.*

(ii) *Every treatment of $G_{j'}$, the set of j'th associates of θ, occurs $p_{ij}^{j'}$ times ($j' = 1, 2, \ldots, m$), in $G_{i,j}$.*

From (8.20) we have

$$\frac{a}{k}\hat{\tau}_i - \sum_{j=1}^{m} \frac{\lambda_j}{k} S_j(\hat{\tau}_i) = Q_i. \tag{8.21}$$

Summing (8.21) over the sth associates of each treatment, using Lemma 8.1 and imposing the restriction $\hat{\tau}_i + \sum_{j=1}^{m} S_j(\hat{\tau}_i) = 0$, we have

$$\sum_{j=1}^{m} a_{sj} S_j(\hat{\tau}_i) = k S_s(Q_i), \quad s = 1, 2, \ldots, m, \tag{8.22}$$

where

$$a_{sj} = \lambda_s n_s - \sum_{\ell=1}^{m} \lambda_\ell p_{s\ell}^j; \quad s \neq j, \ s, j = 1, 2, \ldots, m$$

$$a_{ss} = a + \lambda_s n_s - \sum_{\ell=1}^{m} \lambda_\ell p_{s\ell}^s, \quad s = 1, 2, \ldots, m.$$

Equation (8.22) will be solved for $S_j(\hat{\tau}_i)$ and finally $\hat{\tau}_i$ will be calculated from the relation

$$\hat{\tau}_i = -\sum_{j=1}^{m} S_j(\hat{\tau}_i). \tag{8.23}$$

For a 2-associate class PBIB design, defining the constants, Δ, c_1, c_2, from the relations

$$k^2\Delta = (a + \lambda_1)(a + \lambda_2) + (\lambda_1 - \lambda_2)\left\{a\left(p_{12}^1 - p_{12}^2\right) + \lambda_2 p_{12}^1 - \lambda_1 p_{12}^2\right\},$$

$$k\Delta c_1 = \lambda_1(a + \lambda_2) + (\lambda_1 - \lambda_2)\left(\lambda_2 p_{12}^1 - \lambda_1 p_{12}^2\right),$$

$$k\Delta c_2 = \lambda_2(a + \lambda_1) + (\lambda_1 - \lambda_2)\left(\lambda_2 p_{12}^1 - \lambda_1 p_{12}^2\right),$$

we get

$$\hat{\tau}_i = \frac{k - c_2}{a} Q_i + \frac{c_1 - c_2}{a} S_1(Q_i), \quad i = 1, 2, \ldots, v,$$

which can be written as

$$\hat{\tau} = \left(\frac{k - c_2}{a} B_0 + \frac{c_1 - c_2}{a} B_1\right) \mathbf{Q}_{\tau|\beta}. \tag{8.24}$$

If $\ell'\tau$ is a contrast of treatment effects, its blue is $\ell'\hat{\tau}$ where $\hat{\tau}$ is given by (8.24). Noting that $\sigma^2\left(\frac{k-c_2}{a} B_0 + \frac{c_1-c_2}{a} B_1\right)$ can be treated as the dispersion matrix of $\hat{\tau}$ in finding the variances of blue's of estimable functions of treatment effects, we have

$$V(\hat{\tau}_i - \hat{\tau}_j) = \begin{cases} \dfrac{k - c_1}{k - 1}\left(\dfrac{2\sigma^2}{r}\right), & \text{if } (i, j) = 1; \\[4mm] \dfrac{k - c_2}{k - 1}\left(\dfrac{2\sigma^2}{r}\right), & \text{if } (i, j) = 2. \end{cases}$$

The average variance of all elementary treatment contrasts is then given by

$$\bar{V} = \frac{2\sigma^2}{r} \frac{\{n_1(k - c_1) + n_2(k - c_2)\}}{(v - 1)(k - 1)}, \tag{8.25}$$

and its efficiency E is given by

$$E = \frac{\bar{V}_r}{\bar{V}} = \frac{(k - 1)(v - 1)}{n_1(k - c_1) + n_2(k - c_2)}. \tag{8.26}$$

Clatworthy (1973) in his tables had listed Δ, c_1, c_2 values along with the parameters and plans of the designs. The ANOVA can be completed as described in Sec. 2.2 and a numerical example of the statistical analysis can be found in Cochran and Cox (1957, pp. 456–460).

For a 3-associate class PBIB design, defining

$$A_{13} = r(k-1) + \lambda_3, \quad B_{13} = \lambda_3 - \lambda_1, \quad C_{13} = \lambda_3 - \lambda_2,$$
$$A_{23} = (\lambda_3 - \lambda_1)(n_1 - p_{11}^3) - (\lambda_3 - \lambda_2)p_{12}^3,$$
$$A_{33} = (\lambda_3 - \lambda_2)(n_2 - p_{22}^3) - (\lambda_3 - \lambda_1)p_{12}^3,$$
$$B_{23} = r(k-1) + \lambda_3 + (\lambda_3 - 1)(p_{11}^1 - p_{11}^3) + (\lambda_3 - \lambda_2)(p_{12}^1 - p_{12}^3),$$
$$B_{33} = (\lambda_3 - \lambda_1)(p_{12}^1 - p_{12}^3) + (\lambda_3 - \lambda_2)(p_{12}^1 - p_{22}^3),$$
$$C_{23} = (\lambda_3 - \lambda_1)(p_{11}^2 - p_{11}^3) + (\lambda_3 - \lambda_2)(p_{12}^2 - p_{12}^3),$$
$$C_{33} = r(k-1) + \lambda_3 + (\lambda_3 - \lambda_1)(p_{12}^2 - p_{12}^3) + (\lambda_3 - \lambda_2)(p_{22}^2 - p_{22}^3),$$
$$F = B_{23}C_{33} - B_{33}C_{23},$$
$$G = B_{33}C_{13} - B_{13}C_{33},$$
$$H = B_{13}C_{23} - B_{23}C_{13},$$
$$\Delta_1 = A_{13}F + A_{23}G + A_{33}H,$$

C.R. Rao (1947) showed that Eq. (8.23) simplifies to

$$\hat{\tau}_i = \frac{k}{\Delta_1}\{FQ_i + GS_1(Q_i) + HS_2(Q_i)\}, \quad i = 1, 2, \ldots, v,$$

which can be written as

$$\hat{\tau} = \frac{k}{\Delta_1}\{FB_0 + GB_1 + HB_2\}\mathbf{Q}_{\tau|\beta}.$$

The variances of estimated elementary contrasts of treatment effects can be verified to be

$$V(\hat{\tau}_i - \hat{\tau}_j) = \begin{cases} 2k\sigma^2(F-G)/\Delta_1, & \text{if } (i,j) = 1; \\ 2k\sigma^2(F-H)/\Delta_1, & \text{if } (i,j) = 2; \\ 2k\sigma^2 F/\Delta_1, & \text{if } (i,j) = 3; \end{cases} \tag{8.27}$$

and the efficiency can be verified to be

$$E = \frac{(v-1)\Delta_1}{[(v-1)F - n_1 G - n_2 H]rk}. \tag{8.28}$$

The normal equations (8.20) can also be solved by using the spectral decomposition of $C_{\tau|\beta}$ as

$$C_{\tau|\beta} = \sum_{i=1}^{t} \mu_i A_i, \tag{8.29}$$

where μ_i are the nonzero, distinct, eigenvalues of $C_{\tau|\beta}$ with the corresponding orthogonal idempotent matrices A_i and getting

$$\hat{\tau} = C_{\tau|\beta}^{-} \mathbf{Q}_{\tau|\beta} = \left(\sum_{i=1}^{t} \mu_i^{-1} A_i \right) \mathbf{Q}_{\tau|\beta}. \tag{8.30}$$

This approach was used in obtaining the solutions as (8.30) by some workers including Aggarwal (1973, 1974), Raghavarao (1963), Raghavarao and Chandrasekhararao (1964), and Tharare (1965).

8.4 Some Combinatorics and Constructions

Singular GD designs have $r - \lambda_1 = 0$, and hence every symbol occurs with all its first associates. We thus have

Theorem 8.2 *A singular GD design with parameters $v = mn, b, r, k, \lambda_1, \lambda_2$ and a BIB design with parameters $v_1 = m, b_1 = b, r_1 = r, k_1 = k/n, \lambda = \lambda_2$ coexist. Also, the set size k of a singular GD design is divisible by n.*

By calculating the sum of squares for the number of symbols occurring from each group of the association scheme in the b sets, Bose and Connor (1952) proved the following theorem:

Theorem 8.3 *For a semi-regular GD design, k is divisible by m, and if $k = cm$, then every set of the design has c symbols from each of the m groups of the association scheme.*

Similar to Theorem 8.3, Raghavarao (1960a) proved the following two theorems:

Theorem 8.4 *If in a triangular design,*

$$r + (n - 4)\lambda_1 - (n - 3)\lambda_2 = 0,$$

then $2k$ is divisible by n and each set of the design contains $2k/n$ symbols from each of the n rows of the association scheme.

Theorem 8.5 *If in a L_2 design*

$$r + (s - 2)\lambda_1 - (s - 1)\lambda_2 = 0$$

then k is divisible by s and every set of the design contains k/s symbols from each of the s rows (columns) of the association scheme.

More results on the set structure of these designs are given in Raghavarao (1971). We will now consider some construction methods of these designs:

Theorem 8.6 (*Bush, 1977b*) *A semi-regular group divisible design with parameters*

$$v = 4t - 2, \quad b = 4t, \quad r = 2t, \quad k = 2t - 1, \quad m = 2t - 1,$$
$$n = 2, \quad \lambda_1 = 0, \quad \lambda_2 = t \tag{8.31}$$

exists whenever a BIBD with parameters $v_1 = 4t - 1 = b_1$, $r_1 = 2t - 1 = k_1$, $\lambda_1 = t - 1$ exists.

Proof. Let N^* be the incidence matrix of the BIB design. Choose any $2t - 1$ rows of N^* and number them $1, 2, \ldots, 2t - 1$. Using these numbered rows form $4t - 1$ sets by writing the symbol i or $2t - 1 + i$ in the jth set corresponding to the jth column of N^* according as 1 or 0 occurs in that column. Add a new set by writing the symbols $1, 2, \ldots, 2t - 1$ to get the required design.

Illustration 8.4 Let N^* be the incidence matrix of the BIB design with parameters $v_1 = 7 = b_1, r_1 = 3 = k_1, \lambda_1 = 1$ as given below:

$$N^* = \begin{pmatrix} 1 & 0 & 0 & 0 & 1 & 0 & 1 \\ 1 & 1 & 0 & 0 & 0 & 1 & 0 \\ 0 & 1 & 1 & 0 & 0 & 0 & 1 \\ 1 & 0 & 1 & 1 & 0 & 0 & 0 \\ 0 & 1 & 0 & 1 & 1 & 0 & 0 \\ 0 & 0 & 1 & 0 & 1 & 1 & 0 \\ 0 & 0 & 0 & 1 & 0 & 1 & 1 \end{pmatrix}. \tag{8.32}$$

Choose rows 2, 3, 5 and call them 1, 2, 3. The sets

$$(1, 5, 6); \quad (1, 2, 3); \quad (4, 2, 6); \quad (4, 5, 3); \quad (4, 5, 3);$$
$$(1, 5, 6); \quad (4, 2, 6); \quad (1, 2, 3), \tag{8.33}$$

form a semi-regular group divisible design with parameters

$$v = 6, \quad b = 8, \quad r = 4, \quad k = 3, \quad m = 3, \quad n = 2, \quad \lambda_1 = 0, \quad \lambda_2 = 2, \tag{8.34}$$

with association scheme given by the following rows:

$$\begin{array}{cc} 1 & 4 \\ 2 & 5 \\ 3 & 6 \end{array}$$

Again by choosing rows 2, 3, 7 of N^* and calling them 1, 2, 3 we form the sets

$$\begin{array}{llll}
(1, 5, 6); & (1, 2, 6); & (4, 2, 6); & (4, 5, 3); \\
(4, 5, 6); & (1, 5, 3); & (4, 2, 3); & (1, 2, 3);
\end{array} \tag{8.35}$$

which again forms the semi-regular group divisible design with parameters (8.34). Note that the two solutions (8.33) and (8.35) are not isomorphic.

Theorem 8.7 (*Freeman*, 1976b; *Bush*, 1979) *The* 3 *families of regular group divisible designs with parameters*

$$F_1 : v = 3n = b, r = n + 1 = k, m = 3, n = n, \lambda_1 = n, \lambda_2 = 1; \tag{8.36}$$

$$F_2 : v = 4n = b, r = n + 2 = k, m = 4, n = n, \lambda_1 = n - 2, \lambda_2 = 2; \tag{8.37}$$

$$F_3 : v = 5n, b = 10n, r = 2(n + 1), k = n + 1, m = 5, n = n,$$

$$\lambda_1 = 2n, \lambda_2 = 1; \tag{8.38}$$

always exist.

Proof. Proof is by construction. The incidence matrices of the 3 families of designs are respectively

$$N_1 = \begin{pmatrix} J_n & 0 & I_n \\ I_n & J_n & 0 \\ 0 & I_n & J_n \end{pmatrix},$$

$$N_2 = \begin{pmatrix} J_n - I_n & I_n & I_n & I_n \\ I_n & J_n - I_n & I_n & I_n \\ I_n & I_n & J_n - I_n & I_n \\ I_n & I_n & I_n & J_n - I_n \end{pmatrix},$$

$$N_3 = \begin{pmatrix} J_n & J_n & I_n & I_n & 0 & 0 & 0 & 0 & 0 & 0 \\ I_n & 0 & 0 & 0 & I_n & J_n & J_n & 0 & 0 & 0 \\ 0 & I_n & 0 & 0 & J_n & 0 & 0 & J_n & I_n & 0 \\ 0 & 0 & J_n & 0 & 0 & I_n & 0 & I_n & 0 & J_n \\ 0 & 0 & 0 & J_n & 0 & 0 & I_n & 0 & J_n & I_n \end{pmatrix},$$

where 0 is a zero matrix of appropriate order.

We introduce

Definition 8.12 Given two sets S and T, the Boolean sum of sets S and T, denoted by $S + T$, is

$$S + T = \{x | x \in S \cup T, x \notin S \cap T\}. \tag{8.39}$$

Raghavarao (1974) observed that Boolean sums of sets of PBIB designs, which are linked block designs, are PBIB designs. The Boolean sum of each of the groups of the association scheme with each of the sets of a semi-regular group divisible design will result in a group divisible design.

Let M be a module of m elements $0, 1, \ldots, m-1$ and to each element u of the module, we attach n symbols by writing $u_1, u_2, \ldots, u_n$. We will follow the notation of Sec. 4.6. The following theorems can easily be established.

Theorem 8.8 *Let there exist h initial sets $S_1, S_2, \ldots, S_h$ each of size k in the mn symbols such that*

(i) *each class is represented equally often,*

(ii) *each nonzero pure and mixed differences occurs λ_2 times and each zero mixed difference occurs λ_1 times.*

Then the sets $S_{i\theta} = \theta + S_i$, for $i = 1, 2, \ldots, h; \theta = 0, 1, \ldots, m-1$ form a GD design with parameters $v = mn, b = hm, r = kh/n, k, \lambda_1, \lambda_2$ with the groups of the association scheme given by the m rows

$$
\begin{array}{cccc}
0_1 & 0_2 & \ldots & 0_n \\
1_1 & 1_2 & \ldots & 1_n \\
\vdots & \vdots & \ldots & \vdots \\
(m-1)_1 & (m-1)_2 & \ldots & (m-1)_n
\end{array}
\tag{8.40}
$$

Theorem 8.9 *(Raghavarao and Aggarwal, 1974) Let there exist h initial sets $S_1, S_2, \ldots, S_h$ each of size k in the mn symbols such that*

(i) *each class is represented equally often,*

(ii) *every zero mixed difference occurs λ_1 times, every nonzero pure difference occurs λ_2 times, and every nonzero mixed difference occurs λ_3 times.*

Then the sets $S_{i\theta} = \theta + S_i$ for $i = 1, 2, \ldots, h; \theta = 0, 1, \ldots, m;$ form a rectangular design with parameters $v = mn, b = mh, r = kh/n, k, \lambda_1, \lambda_2,$ and λ_3 with association scheme consisting of the m rows and n columns of (8.40).

Theorem 8.10 *(Raghavarao, 1973) Let $m = n = s$ and let it be possible to find initial sets $S_1, S_2 \ldots, S_h$ such that*

(i) *each class is represented equally often,*

(ii) *every nonzero pure difference and every zero mixed difference occurs λ_1 times, and every nonzero mixed difference occurs λ_2 times.*

Then the sets $S_{i\theta} = \theta + S_i$ for $i = 1, 2, \ldots, h; \theta = 0, 1, \ldots, s-1$ form a L_2 design with parameters $v = s^2, b = sh, r = kh/s, k, \lambda_1, \lambda_2$ with the association scheme (8.40) where $m = n = s$.

Illustration 8.5 The 7 initial sets

$$(0_{1+i}, 1_{1+i}, 3_{1+i}, 0_{2+i}, 1_{2+i}, 3_{2+i}, 0_{4+i}, 1_{4+i}, 3_{4+i}), \quad i = 0, 1, \ldots, 6, \quad (8.41)$$

where the subscripts of the elements in the sets of (8.41) are taken mod 7, when developed, mod 7 gives the solution of the L_2 design with parameters

$$v = 49 = b, \quad r = 9 = k, \quad \lambda_1 = 3, \quad \lambda_2 = 1,$$

the parametric combination which was mistakenly listed as unsolved at number 4 in Table 8.10.1 of Raghavarao (1971). The above solution is isomorphic to the solutions given by Archbold and Johnson (1956) and Vartak (1955). The later isomorphism was demonstrated by John (1975).

By using projective geometries, Raghavarao (1971) indicated the following two theorems:

Theorem 8.11 *If s is a prime or prime power, a semi-regular GD design with the parameters*

$$v = ms, \quad b = s^3, \quad r = s^2, \quad k = m, \quad n = s, \quad \lambda_1 = 0, \quad \lambda_2 = s, \quad (8.42)$$

where $m \leq s^2 + s + 1$, always exists.

Theorem 8.12 *If s is a prime or prime power, a regular GD design with the parameters*

$$v = b = s(s - 1)(s^2 + s + 1), \quad r = k = s^2, \quad m = s^2 + s + 1,$$

$$n = s(s - 1), \quad \lambda_1 = 0, \quad \lambda_2 = 1 \quad (8.43)$$

always exist.

In $\mathrm{PG}(t, s)$ by taking certain subset of the points as symbols and certain flats as sets, a class of PBIB designs with the parameters

$$
\begin{aligned}
v &= (s^{t+1} - s^{\pi+1})/(s - 1), \quad k = (s^{\mu+1} - s^{\upsilon+1})/(s - 1), \\
b &= s^{(\pi-\upsilon)(\mu-\upsilon)}\phi(t - \pi - 1, \mu - \upsilon - 1, s)\phi(\pi, \upsilon, s), \\
r &= s^{(\pi-\upsilon)(\mu-\upsilon-1)}\phi(t - \pi - 2, \mu - \upsilon - 2, s)\phi(\pi, \upsilon, s), \\
m &= s^{(t-\pi)}/(s - 1), \quad n = s^{\pi+1}, \\
\lambda_1 &= s^{(\pi-\upsilon)(\mu-\upsilon-1)}\phi(t - \pi - 2, \mu - \upsilon - 2, s)\phi(\pi - 1, \upsilon - 1, s), \\
\lambda_2 &= s^{(\pi-\upsilon)(\mu-\upsilon-2)}\phi(t - \pi - 3, \mu - \upsilon - 3, s)\phi(\pi, \upsilon, s),
\end{aligned}
\quad (8.44)
$$

for integers, t, μ, υ and π (≥ 0) such that $-1 \leq \upsilon \leq \pi < t - 1$ and $\pi + \mu - t \leq \upsilon < \mu < t$ was constructed by Hamada (1974).

Hamada and Tamari (1975) introduced affine geometrical association scheme and showed that the duals of BIB designs obtained by taking μ-flats of $\mathrm{EG}(t, s)$

are PBIB designs with affine geometrical association scheme in $m = \min\{2\mu + 1, 2(t - \mu)\}$ associate classes.

Ralston (1978) constructed the following series of L_{s-3} design with the help of projective geometry:

$$v = s^2 = b, \quad r = s - 1 = k, \quad n_1 = (s - 2)(s - 1),$$
$$n_2 = 3(s - 1), \quad \lambda_1 = 1, \quad \lambda_2 = 0. \tag{8.45}$$

Let 2 BIB designs exist with parameters v_i, b_i, r_i, k_i and λ_i with incidence matrices N_i for $i = 1, 2$ and let one of them satisfy the condition $b_i = 4(r_i - \lambda_i)$. In N_i, replace 1 by N_{3-i} and 0 by $\bar{N}_{3-i}$, the complement of N_{3-i}. Then Trivedi and Sharma (1975) showed that we get a group divisible design with parameters

$$v = v_i v_{3-i}, \quad b = b_i b_{3-i}, \quad r = r_i r_{3-i} + (b_i - r_i)(b_{3-i} - r_{3-i}),$$
$$k = k_i k_{3-i} + (v_i - k_i)(v_{3-i} - k_{3-i}),$$
$$\lambda_1^* = r_i \lambda_{3-i} + (b_{3-i} - 2r_{3-i} + \lambda_{3-i})(b_i - r_i), \tag{8.46}$$
$$\lambda_2^* = r_{3-i} \lambda_i + (b_i - 2r_i + \lambda_i)(b_{3-i} - r_{3-i}),$$
$$m = v_{3-i}, \quad n = v_i.$$

18 sets of parametric combinations of triangular designs were given in Table 8.8.1 of Raghavarao (1971), whose solutions are unknown, and Aggarwal (1972) found the solution, by trial and error, of the parametric combination listed as number 7 in that table (also see Nigam 1974a).

The L_2 design listed as number 2 in Table 8.10.1 of Raghavarao (1971) having parameters

$$v = 36, \quad b = 60, \quad r = 10, \quad k = 6, \quad \lambda_1 = \lambda_2 = 2 \tag{8.47}$$

was obtained by Sharma (1978) (see also Stahly, 1976) as a particular case of the following:

Theorem 8.13 *An L_2 design with the parameters*

$$v = s^2, \quad b = 2s(s - 1), \quad r = 2(s - 1), \quad k = s, \quad \lambda_1 = 0, \quad \lambda_2 = 2 \tag{8.48}$$

can always be constructed when it is possible to arrange the symbols within each set of the irreducible BIB design with parameters

$$v_1 = s = b_1, \quad r_1 = s - 1 = k_1, \quad \lambda_1^* = s - 2$$

and with symbols $s, s + 1, \ldots, 2s - 1$ in such a way that each distinct pair of symbols θ and ϕ are adjacent in one set, one symbol apart in one set, two symbols apart in one set, $\ldots$, and $(s - 3)$ symbols apart in one set, when the sets are considered as cycles.

The proof is basically the method of construction. Before we describe the construction, we will illustrate the condition required on the irreducible design. The sets

$$(6, 7, 9, 8); \quad (7, 8, 5, 9); \quad (8, 9, 6, 5);$$
$$(9, 5, 7, 6); \quad (5, 6, 8, 7)$$

of the irreducible BIB design with parameters

$$v_1 = 5 = b_1, \quad r_1 = 4 = k_1, \quad \lambda_1^* = 3$$

satisfies the condition. The symbols, say 6 and 8, appear adjacent in set 5; one symbol apart in set 3; two symbols apart in set 1.

When s is a prime or prime power, the initial set $(x^0, x, x^2, \ldots, x^{s-2})$ developed mod s and s added to each element of each set provides the solution with the needed condition, where x is a primitive root of GF(s).

Proof. Let $v = s^2$ symbols be arranged in an $s \times s$ square array S

$$
\begin{array}{ccccc}
0 & 1 & 2 & \cdots & s-1 \\
s & s+1 & s+2 & \cdots & 2s-1 \\
\vdots & \vdots & \vdots & \cdots & \vdots \\
s(s-1) & s(s-1)+1 & s(s-1)+2 & \cdots & s^2-1
\end{array}
$$

Omit the first row of S to get an $(s-1) \times s$ rectangular array R. Omit the ith column of R and rearrange the columns so that its first row is identical with the ith set of the irreducible BIB design satisfying the condition on the theorem. We call such a square R_i^*. Superimpose the Latin square

$$
\begin{array}{ccccc}
\alpha_1 & \alpha_2 & \alpha_3 & \cdots & \alpha_{s-1} \\
\alpha_{s-1} & \alpha_1 & \alpha_2 & \cdots & \alpha_{s-2} \\
\alpha_{s-2} & \alpha_{s-1} & \alpha_1 & \cdots & \alpha_{s-3} \\
\vdots & \vdots & \vdots & \vdots & \vdots \\
\alpha_2 & \alpha_3 & \alpha_4 & \cdots & \alpha_1
\end{array}
$$

and form $s - 1$ sets of the symbols of R_i^* corresponding to each symbol of the above Latin square and to each of the $s - 1$ sets augment the symbol i, so that the sets formed are of size s. Again on R_i^* superimpose the Latin square.

$$
\begin{array}{ccccc}
\alpha_2 & \alpha_3 & \cdots & \alpha_{s-1} & \alpha_1 \\
\alpha_3 & \alpha_4 & \cdots & \alpha_1 & \alpha_2 \\
\alpha_4 & \alpha_5 & \cdots & \alpha_2 & \alpha_3 \\
\vdots & \vdots & \vdots & \vdots & \vdots \\
\alpha_{s-1} & \alpha_1 & \cdots & \alpha_{s-3} & \alpha_{s-2} \\
\alpha_1 & \alpha_2 & \cdots & \alpha_{s-2} & \alpha_{s-1}
\end{array}
$$

and form $s - 1$ sets as before. Thus from R_i^* we generate $2(s - 1)$ sets of size s. By repeating this process for each $i = 0, 1, \ldots, s - 1$ we generate $2s(s - 1)$ sets forming the required L_2 design.

Two methods of combining two given PBIB designs with m and n associate classes respectively to obtain new PBIB designs with (i) $m + n$ associate classes, and (ii) $m + n + mn$ associate classes were discussed by Saha (1978).

8.5 Partial Geometric Designs

Generalizing the idea of partial geometry (r, k, t) (see Raghavarao, 1971; Chap. 9), Bose, Shrikhande and Singhi (1976) introduced partial geometric designs (r, k, t, c). Let v symbols be arranged in b sets such that each set contains k distinct symbols and each symbol occurs in r sets. Let $N = (n_{ij})$ be the incidence matrix and we assume the design to be connected. Let $\lambda(x_i, x)$ be the number of sets in which symbols x_i and x occur together. Let

$$
m(x_i, S_j) = \sum_{x_\alpha \in S_j} \lambda(x_i, x_\alpha) = \sum_{\alpha=1}^{v} \lambda(x_i, x_\alpha) n_{\alpha j}. \tag{8.49}
$$

We then have the following:

Definition 8.13 The configuration (v, b, r, k) is said to be a *partial geometric design* (r, k, t, c) for $t \geq 1$ if

$$
m(x_i, S_j) = \begin{cases} t, & \text{if } x_i \notin S_j; \\ r + k - 1 + c, & \text{if } x_i \in S_j. \end{cases} \tag{8.50}
$$

Using (8.49) and Definition 8.13, we get

Theorem 8.14 *A necessary and sufficient condition that a (v, b, r, k) configuration with incidence matrix N is a partial geometric design (r, k, t, c) is that*

$$
NN'N = (r + k + c - 1 - t)N + t J_{vb}. \tag{8.51}
$$

From (8.51) we get

$$NN'(N\mathbf{1}_b) = \left(\frac{bt}{r} + r + k + c - 1 - t\right) r\mathbf{1}_v,$$

from which it follows that

$$NN'\mathbf{1}_v = \left(\frac{bt}{r} + r + k + c - 1 - t\right)\mathbf{1}_v.$$

Again $NN'\mathbf{1}_v = rk\mathbf{1}_v$. Hence we have

$$\left(\frac{bt}{r} + r + k + c - 1 - t\right) = rk. \tag{8.52}$$

Let $\boldsymbol{\xi}_i$ be a b-component vector such that $\boldsymbol{\xi}_i'\mathbf{1}_b = 0$ and $\boldsymbol{\xi}_i \neq 0$. There exist Rank $(N) - 1 = \alpha$ (say) such $\boldsymbol{\xi}_i$'s and

$$NN'N\boldsymbol{\xi}_i = (r + k + c - 1 - t)(N\boldsymbol{\xi}_i). \tag{8.53}$$

Thus $\theta = r + k + c - 1 - t$ is an eigenvalue of multiplicity α of NN'. α can be evaluated from the relation

$$\alpha(r + k + c - 1 - t) + rk = vr,$$

from which we get

$$\alpha = \frac{r(v - k)}{r + k + c - 1 - t}. \tag{8.54}$$

Thus

Theorem 8.15 (*Bose, Bridges and Shrikhande*, 1976) *NN' of a partial geometric design (r, k, t, c) has only one nonzero eigenvalue $\theta = r + k + c - 1 - t$ other than the simple root rk and its multiplicity α is given by* (8.54).

Bose and Shrikhande (1979) showed that all PBIB designs with 2 associate classes and $r < k$ are necessarily partial geometric designs, generalizing Theorem 9.6.2 of Raghavarao (1971).

8.6 Applications to Group Testing Experiments

We will consider hypergeometric, non-adaptive group testing designs to identify 2 defective items among n items in t tests. Let $S_1, S_2, \ldots, S_t$ be the tests for group testing, which can be considered as sets of a block design in $v = n$ symbols (items). We avoid the trivial case of one item per test. Let $S_1^*, S_2^*, \ldots, S_n^*$ be the dual design in t symbols. Weideman and Raghavarao (1987a) noted that the original design identifies 2 defective items if

$$P_1 : S_i^* \cup S_j^* \neq S_{i'}^* \cup S_{j'}^*, \quad i, j, i', j' = 1, 2, \ldots, n; \quad \{i, j\} \neq (i', j').$$

The property P_1 was given in a different form by Hwang and Sos (1981). To systematically study the designs, Weideman and Raghavarao (1987a) imposed another condition P_2 given by

P_2: In $S_1^*, S_2^*, \ldots, S_n^*$ every pair of symbols occurs at most once.

Violation of P_2 does not destroy the identifiablilty property of defective items. They noted that the dual design can always be constructed satisfying P_1 and P_2 with sets of size 2 or 3. In fact if S_i^* has more than 3 symbols, any subset of 3 elements of S_i^* can be used in place of S_i^*. They proved

Theorem 8.16 (*Weideman and Raghavarao*, 1987a) *The number of items, n, to be tested in t group tests, where the dual design satisfies P_1 and P_2 and has set size of 2 or 3 satisfies*

$$n \leq \lfloor t(t+1)/6 \rfloor, \tag{8.55}$$

where $\lfloor \bullet \rfloor$ is the greatest integer function.

The designs for $t \equiv 0, 2, 4 \pmod{6}$ can be constructed using GD designs. When $t \equiv 0$ or $2 \pmod{6}$, the dual design consists of a GD design with parameters $v = t$, $b = t(t-2)/6, k = 3, r = (t-2)/2, m = t/2, n = 2, \lambda_1 = 0, \lambda_2 = 1$, if it exists, and sets formed by writing $t/2$ groups of the association scheme.

Illustration 8.6 Let $t = 12$, so that $n \leq 26$. We consider the GD design with parameters $v = 12, b = 20, r = 5, k = 3, m = 6, n = 2, \lambda_1 = 0, \lambda_2 = 1$ given as $R70$ in Clatworthy's Tables, and the association scheme to form the sets of the dual design as follows:

$$(1, 3, 4); \quad (2, 9, 7); \quad (3, 5, 12); \quad (9, 6, 10); \quad (6, 8, 1);$$
$$(11, 4, 2); \quad (8, 7, 3); \quad (4, 12, 9); \quad (7, 10, 5); \quad (10, 1, 11);$$
$$(2, 5, 6); \quad (3, 6, 11); \quad (9, 11, 8); \quad (5, 8, 4); \quad (6, 4, 7);$$
$$(11, 7, 12); \quad (8, 12, 10); \quad (12, 1, 2); \quad (10, 2, 3); \quad (1, 9, 5);$$
$$(1, 7); \quad (2, 8); \quad (3, 9); \quad (4, 10); \quad (5, 11); \quad (6, 12).$$

By dualizing the above design, we get a group testing design to test 26 items in 12 tests and identify the 2 defective items.

Test #	Item # included in the test
1	1, 5, 10, 18, 20, 21
2	2, 6, 11, 18, 19, 22
3	1, 3, 7, 12, 19, 23
4	1, 6, 8, 14, 15, 24
5	3, 9, 11, 14, 20, 25
6	4, 5, 11, 12, 15, 26
7	2, 7, 9, 15, 16, 21
8	5, 7, 13, 14, 17, 22
9	2, 4, 8, 13, 20, 23
10	4, 9, 10, 17, 19, 24
11	6, 10, 12, 13, 16, 25
12	3, 8, 16, 17, 18, 26

If, for example, items 3 and 20 are defective, tests 1, 3, 5, 9 and 12 will give positive results, and the union of the tests with negative result contains all items except 3 and 20.

In the case $t \equiv 4 \pmod{6}$, Weideman and Raghavarao (1987b) constructed dual designs, by trial and error, for $t = 10$ and 16. For $t \geq 22$, and $t = 6p + 4$, the dual design consists of the solution of a GD design with parameters.

$$v = 6p, \quad b = 6p(p-1), \quad r = 3(p-1), \quad k = 3,$$
$$m = p, \quad n = 6, \quad \lambda_1 = 0, \quad \lambda_2 = 1 \tag{8.56}$$

with association scheme consisting of p groups

$$i, p+i, 2p+i, 3p+i, 4p+i, 5p+i, \quad i = 1, 2, \ldots, p;$$

and the following $15p + 3$ sets:

$$
\begin{array}{ll}
(i, p+i); & (p+i, 5p+i, 6p+1); \\
(i, 2p+i, 6p+1); & (2p+i, 3p+i); \\
(i, 3p+i, 6p+2); & (2p+i, 4p+i, 6p+4); \\
(i, 4p+i, 6p+3); & (2p+i, 5p+i, 6p+2); \\
(i, 5p+i, 6p+4); & (3p+i, 4p+i, 6p+1); \\
(p+i, 2p+i, 6p+3); & (3p+i, 5p+i, 6p+3); \\
(p+i, 3p+i, 6p+4); & (4p+i, 5p+i); \\
(p+i, 4p+i, 6p+2); & (6p+1, 6p+2); \\
(6p+1, 6p+3); & (6p+1, 6p+4),
\end{array}
$$
$$i = 1, 2, \ldots, p. \tag{8.57}$$

The dual design has $6p(p - 1) + 15p + 3 = 3(2p + 1)(p + 1)$ sets in $6p + 4$ symbols.

Illustration 8.7 For $t = 22$, $n \leq \lfloor 84.3 \rfloor = 84$. We construct the dual design consisting of the semi-regular GD design with parameters

$$v = 18, \quad b = 36, \quad r = 6, \quad k = 3, \quad m = 3, \quad n = 6, \quad \lambda_1 = 0, \quad \lambda_2 = 1$$

which is listed as SR 30 in Clatworthy's Tables and the 48 sets (8.57) for $p = 3$. The required dual design is

$$(1, 2, 3); (1, 5, 6); (8, 9, 1); (11, 1, 12); (14, 1, 15);$$
$$(17, 18, 1); (4, 2, 18); (5, 4, 3); (4, 8, 6); (11, 4, 9);$$
$$(14, 12, 4); (17, 15, 4); (7, 15, 2); (7, 18, 5); (3, 7, 8);$$
$$(6, 11, 7); (9, 7, 14); (12, 17, 7); (10, 12, 2); (10, 5, 15);$$
$$(18, 10, 8); (3, 10, 11); (6, 14, 10); (9, 17, 10); (13, 3, 14);$$
$$(13, 6, 17); (12, 3, 5); (15, 8, 13); (18, 13, 11); (2, 9, 13);$$
$$(16, 3, 17); (16, 14, 18); (15, 11, 16); (2, 6, 16); (5, 16, 9);$$
$$(8, 16, 12); (1, 4); (2, 5); (3, 6); (1, 7, 19); (2, 8, 19);$$
$$(3, 9, 19); (1, 10, 20); (2, 11, 20); (3, 12, 20); (1, 13, 21);$$
$$(2, 14, 21); (3, 15, 21); (1, 16, 22); (2, 17, 22); (3, 18, 22);$$
$$(4, 7, 21); (5, 8, 21); (6, 9, 21); (4, 10, 22);$$
$$(5, 11, 22); (6, 12, 22); (4, 13, 20); (5, 14, 20);$$
$$(6, 15, 20); (4, 16, 19), (5, 17, 19); (6, 18, 19);$$
$$(7, 10); (8, 11); (9, 12); (7, 13, 22); (8, 14, 22);$$
$$(9, 15, 22); (7, 16, 20); (8, 17, 20); (9, 18, 20);$$
$$(10, 13, 19); (11, 14, 19); (12, 15, 19);$$
$$(10, 16, 21); (11, 17, 21); (12, 18, 21);$$
$$(12, 16); (14, 17); (15, 18); (19, 20); (19, 21); (19, 22).$$

By dualizing the above design, we get the group testing design to test 84 items in 22 tests and identify 2 defective items.

Vakil, Parnes and Raghavarao (1990) considered group testing designs where every item is included in exactly 2 group tests. They proved

Theorem 8.17 *If each item is included in exactly 2 group tests and if there are at most 2 defective items among the n items, then the number of tests t and n satisfy*

$$n \leq \lfloor t\sqrt{t - 1}/2 \rfloor. \tag{8.58}$$

When $t - 1$ is a perfect square, they noted that the dual of the group testing design referenced in Theorem 8.17 is a simple PBIB design with parameters

$$v = t, \quad b = n, \quad r = \sqrt{t-1}, \quad k = 2, \quad \lambda_1 = 1, \quad \lambda_2 = 0,$$
$$n_1 = \sqrt{t-1}, \quad n_2 = \sqrt{t-1}(\sqrt{t-1}-1), \quad p_{11}^1 = 0. \tag{8.59}$$

8.7 Applications in Sampling

While sampling from finite populations, it is often convenient to form strata of homogeneous units. We consider the strata of equal sizes. Let the population consist of M strata of N units and let $\bar{Y}_i$ and S_i^2 be the ith stratum population mean and variance for $i = 1, 2, \ldots, M$. Let $\bar{Y} = \sum_{i=1}^{M} \bar{Y}_i / M$. We want to take a sample of size n with proportional allocation, which in this case is equal allocation. Let $n = Mc$, where $c \geq 2$.

Suppose a semi-regular GD design exists with parameters

$$v = MN, \quad b, r, k = n = Mc, \lambda_1, \lambda_2$$

with M groups (strata) of N symbols (units). We select a set of this semi-regular GD with equal probability as our sample. Let $\bar{y}$ be the sample mean and s_i^2 be the sample variance from the sample units of the ith straum ($i = 1, 2, \ldots, M$).

Following a similar argument as in Sec. 5.1 (also see Raghavarao and Singh, 1975), we have the following theorem.

Theorem 8.18 *$\bar{y}$ is an unbiased estimator of $\bar{Y}$ and*

$$\mathrm{var}(\bar{y}) = \frac{v-n}{vn} \sum_i S_i^2. \tag{8.60}$$

Furthermore,

$$\hat{\mathrm{var}}(\bar{y}) = \frac{v-n}{vn} \sum_i s_i^2. \tag{8.61}$$

Raghavarao and Singh (1975) also discussed the use of L_2 and rectangular designs in cluster sampling.

8.8 Applications in Intercropping

In intercropping experiments, sometimes the farmers use a primary crop as a revenue generator and mix with several other secondary crops (see Raghavarao and Rao, 2001). Some of the secondary crops are resistant to natural damages like 1. heavy rains, 2. too much heat, 3. pests, 4. disease, etc.

Suppose the farmer wants to use m categories of secondary crop by choosing one from each category and let us assume that n varieties are available in each category. Then the problem is to choose the best variety in each category to form the secondary crops mix to be used with the primary crop.

Suppose a semi-regular GD design exists with parameters

$$v = mn, \quad b = n^2, \quad r = n, \quad k = m, \quad \lambda_1 = 0, \quad \lambda_2 = 1, \tag{8.62}$$

we use the blend of varieties in each set of the design with parameters (8.62) as secondary crops to be sown together with the primary crop. Let the response of yield on the primary crop be measured with each blend. We denote the jth variety in the ℓth category as (i, j) for $j = 1, 2, \ldots, n; i = 1, 2, \ldots, m$.

Let $S_\alpha = \{(1, i_1), (2, i_2), \ldots, (m, i_m)\}$ be the αth set of the GD design (8.62). The response Y_α on the primary crop with the blend S_α can then be modeled as

$$E(Y_\alpha) = \tau + \sum_{\ell=1}^{m} \beta_{\ell i_\ell}, \tag{8.63}$$

where τ is the effect of the main crop and $\beta_{\ell i_\ell}$ is the competition effect of the i_ℓ variety in the ℓth category on the main crop. We want to draw inferences on $\beta_{\ell i_\ell} - \beta_{\ell j_\ell}$, where i_ℓ and j_ℓ are two different varieties in the ℓth category.

Let us reparametrize the competing effects so that $\sum_{j=1}^{n} \beta_{ij} = 0$ for every $i = 1, 2, \ldots, m$, and change τ to τ^* so that

$$E(Y_\alpha) = \tau^* + \sum_{\ell=1}^{m} \beta_{\ell i_\ell}, \tag{8.64}$$

where $\sum_{j=1}^{n} \beta_{ij} = 0$, for every $i = 1, 2, \ldots, m$. We assume $m < n + 1$. Let

$$T_{ij} = \sum_{\alpha:(i,j)\in S_\alpha} Y_\alpha, \quad G = \sum_{\alpha=1}^{n^2} Y_\alpha \tag{8.65}$$

and

$$\bar{T}_{ij} = T_{ij}/n. \tag{8.66}$$

Assuming that Y_α's are independently normally distributed with variance σ^2 and using the results of Chap. 1, the minimum residual sum of squares, R_0^2 is given by

$$R_0^2 = \left\{ \sum_{\alpha=1}^{n^2} Y_\alpha^2 - \frac{G^2}{n^2} \right\} - \left\{ \sum_{i=1}^{m} \sum_{j=1}^{n} \frac{T_{ij}^2}{n} - \frac{G^2}{n^2} \right\} \tag{8.67}$$

with $(n-1)(n+1-m)$ degrees of freedom. Hence the estimate of σ^2 is

$$\hat{\sigma}^2 = R_0^2/\{(n-1)(n+1-m)\}, \tag{8.68}$$

where R_0^2 is given by (8.67). Further

$$\hat{\beta}_{ij} - \hat{\beta}_{ij'} = \bar{T}_{ij} - \bar{T}_{ij'}. \tag{8.69}$$

The null hypothesis $H_0 : \beta_{ij} = \beta_{ij'}$ will be tested using the test statistic

$$t = \frac{\bar{T}_{ij} - \bar{T}_{ij'}}{\sqrt{2\hat{\sigma}^2/n}}, \tag{8.70}$$

which has a t-distribution with $(n-1)(n+1-m)$ degrees of freedom. The p-value for a one-sided or two-sided alternatives can be easily calculated, and conclusions are drawn.

Alternatively, one can use Scheffe's multiple comparison method. The intercrops (i, j) and (i, j') are concluded significant, if

$$|\bar{T}_{ij} - \bar{T}_{ij'}| > \sqrt{m(n-1)F_{1-\alpha}(m(n-1), (n-1)(n+1-m))}\sqrt{2\hat{\sigma}^2/r}, \tag{8.71}$$

where $F_{1-\alpha}(m(n-1), (n-1)(n+1-m))$ is the $100(1-\alpha)$ percentile point of an F-distribution with $m(n-1)$ numerator, and $(n-1)(n+1-m)$ denominator degrees of freedom.

8.9 Concluding Remarks

The triangular association scheme is uniquely defined by the parameters except when $n = 8$, in which case there are 3 pseudo triangular association schemes (see Chang, 1960). L_2 association scheme is uniquely defined by the parameters except when $s = 4$, in which case there is 1 pseudo L_2 association scheme (see Shrikhande, 1959).

Necessary conditions for the existence of symmetric and some asymmetric PBIB designs using Hasse–Minkowski invariant were well documented in Raghavarao (1971, Chap. 12).

There is a close relationship between PBIB designs and strongly regular graphs in graph theory.

Shah (1959a) relaxed the condition $p_{jj'}^i = p_{j'j}^i$ in PBIB designs and developed a class of block designs whose statistical analysis is similar to the usual PBIB design analysis. Nair (1964) relaxed the condition of symmetry in the relation of association, i.e. if ϕ is an ith associate of θ, then θ is not necessarily the ith associate of ϕ.

Multidimensional partially balanced designs were studied by Srivastava and Anderson (1970).

The concept of linked block designs was generalized by Roy and Laha (1957) and Nair (1966) to partially linked block designs, which are duals of PBIB designs.

By combining two PBIB designs with the same association schemes and different set sizes so that the sum of the two corresponding λ_i parameters are constant, Raghavarao (1962a) obtained combinatorically balanced Symmetrical Unequal Block (SUB) arrangements.

When BIB designs, or LB designs do not exist, the optimal designs are GD designs with $\lambda_2 = \lambda_1 \pm 1$ (see Cheng 1980). Also, see Cheng and Bailey (1991) for the optimality results of 2-associate class PBIB designs with $\lambda_2 = \lambda_1 \pm 1$, which are Regular Graph Designs (RGD) given by John and Mitchell (1977). An equi-replicated, equi-block sized design with parameters v, b, r, k is called RGD if

(i) every treatment occurs $\lfloor k/v \rfloor$ or $\lfloor k/v \rfloor + 1$ times in each block, where $\lfloor \bullet \rfloor$ is the greatest integer function, and

(ii) $|\lambda_{ij} - \lambda_{i'j'}| \leq 1$ where $\lambda_{ij} (\lambda_{i'j'})$ are the number of blocks in which treatments i and j (i' and j') occur together.

For other results on the optimality of RGD, we refer to Shah and Sinha (1989).

9

Lattice Designs

9.1 Introduction

Sometimes in experiments a large number of treatments are to be tested in blocks of small sizes. This is especially the case in plant breeding trials, where the scientists need to select the best varieties from a large number of available varieties. Usually in such experiments, enough material will not be available on each variety to have a large trial and they are interested in designs with 2 or 3 replications. Lattice designs are useful in laying out such trials.

In a square lattice we have $v = s^2$ treatments and it is pth dimensional if the design has p replications. The construction is based upon using $p - 2$ mutually orthogonal Latin squares of order s and we will discuss it in detail in Sec. 9.2. When $p = 2$, the lattice design is called a simple square lattice, and when $p = s + 1$, it is called a balanced lattice. A balanced lattice is an affine resolvable BIB design with parameters.

$$v = s^2, \quad b = s(s+1), \quad r = s+1, \quad k = s, \quad \lambda = 1. \tag{9.1}$$

A cubic lattice design has $v = s^3$ treatments arranged in blocks of size s. A simple cubic lattice design has 3 replications.

A rectangular lattice has $v = s(s-1)$ treatments arranged in a resolvable design of s blocks of size $s - 1$ in each replication. The number of replications used is the dimensionality of the rectangular lattice.

Lattice designs were originally introduced into experimental work by Yates (1940) and rectangular lattices were developed by Harshberger (1947, 1950). The square and cubic lattice designs and some rectangular lattices are PBIB designs. Many rectangular lattices are not PBIB designs and they form a more general class of block designs.

9.2 Square Lattice Designs

We briefly mentioned orthogonal Latin squares in Chap. 4 and we will use them to construct square lattices of different orders.

Recall that a Latin square is an arrangement of s symbols in an $s \times s$ square array such that every symbol occurs exactly once in each row and each column. Given two Latin squares, they are said to be orthogonal if on superimposition of one square on the other, each of the s^2 ordered pairs occurs exactly once. A set of Latin squares, in which each pair is orthogonal, is said to be a set of Mutually Orthogonal Latin Squares (MOLS). It can be shown that the number of MOLS of order s is at most $s - 1$. A set of $s - 1$, MOLS of order s is called a complete set of MOLS and they exist when s is a prime or prime power. A complete set of MOLS of order 4 was displayed on p. 68 of this monograph and is reproduced below

$$
\begin{array}{|cccc|}
\hline
0 & 1 & 2 & 3 \\
1 & 0 & 3 & 2 \\
2 & 3 & 0 & 1 \\
3 & 2 & 1 & 0 \\
\hline
\end{array}
$$

$$
\begin{array}{|cccc|}
\hline
0 & 2 & 3 & 1 \\
1 & 3 & 2 & 0 \\
2 & 0 & 1 & 3 \\
3 & 1 & 0 & 2 \\
\hline
\end{array}
\tag{9.2}
$$

$$
\begin{array}{|cccc|}
\hline
0 & 3 & 1 & 2 \\
1 & 2 & 0 & 3 \\
2 & 1 & 3 & 0 \\
3 & 0 & 2 & 1 \\
\hline
\end{array}
$$

When s is a prime or prime power, let $\alpha_0 = 0$, $\alpha_i = x^i$, $i = 1, 2, \ldots, s - 1$ be the elements of $GF(s)$, where x is a primitive root. Then the Latin square L_t of the complete set of MOLS is constructed by putting $\alpha_i + \alpha_j \alpha_t$ as its (i, j) element for $i, j = 0, 1, \ldots, s - 1$. The set of MOLS (9.2) are constructed by this method and we used i for α_i.

For $s = 10$, two MOLS exist and it is unknown whether three MOLS exist. For order $s = 6$, no pair of MOLS exist. For more details on Latin squares and MOLS, the reader is referred to Denes and Keedwell (1974) or Raghavarao (1971, Chaps. 1 and 3).

We will turn our attention to square lattice design with $v = s^2$ treatments. We arrange all the symbols in an $s \times s$ square array. By writing the rows of the array as the first replication of s blocks and the columns of the array as the second replication of s blocks, we form a simple square lattice design. We superimpose a Latin square on the array and form the third replication of s blocks, where the ith block consists of symbols coinciding with the ith letter of the Latin square. The three replications become a triple square lattice design. We take a Latin square orthogonal to the earlier Latin square and superimpose it on the $s \times s$ array of s^2 symbols. The fourth replication consists of s blocks, where the ith block consists of symbols coinciding with the ith letter of the second Latin square. The four replications constitute a four-dimensional square lattice design. Continuing in this fashion, when s is a prime or prime power, we construct a balanced square lattice, using a complete set of MOLS, consisting of $s + 1$ replications.

Illustration 9.1 We will illustrate the construction with $v = 16$, by writing the 16 symbols 0, 1, ..., 15 in a 4 × 4 array

$$
\begin{array}{cccc}
0 & 1 & 2 & 3 \\
4 & 5 & 6 & 7 \\
8 & 9 & 10 & 11 \\
12 & 13 & 14 & 15
\end{array}
\tag{9.3}
$$

The blocks corresponding to the rows of (9.3) forming the first replication are

$$
\begin{aligned}
&(0, 1, 2, 3), \\
&(4, 5, 6, 7), \\
&(8, 9, 10, 11), \\
&(12, 13, 14, 15).
\end{aligned}
\tag{9.4}
$$

The blocks corresponding to the columns of (9.3) forming the second replication are

$$
\begin{aligned}
&(0, 4, 8, 12), \\
&(1, 5, 9, 13), \\
&(2, 6, 10, 14), \\
&(3, 7, 11, 15).
\end{aligned}
\tag{9.5}
$$

The blocks corresponding to the symbols of the first Latin square of (9.2) forming the third replication are

$$
\begin{aligned}
&(0, 5, 10, 15), \\
&(1, 4, 11, 14), \\
&(2, 7, 8, 13), \\
&(3, 6, 9, 12).
\end{aligned}
\tag{9.6}
$$

The blocks corresponding to the symbols of the second Latin square of (9.2) forming fourth replication are

$$
\begin{aligned}
&(0, 7, 9, 14), \\
&(3, 4, 10, 13), \\
&(1, 6, 8, 15), \\
&(2, 5, 11, 12).
\end{aligned}
\tag{9.7}
$$

The blocks corresponding to the symbols of the last Latin square of (9.2) forming fifth replication are

$$
\begin{aligned}
&(0, 6, 11, 13), \\
&(2, 4, 9, 15), \\
&(3, 5, 8, 14), \\
&(1, 7, 10, 12).
\end{aligned}
\tag{9.8}
$$

The 8 blocks (9.4) and (9.5) form a simple square lattice. The blocks (9.4)–(9.6) form a triple square lattice. The 20 blocks (9.4)–(9.8) form the balanced square lattice, which is a BIB design with parameters

$$
v = 16, \quad b = 20, \quad r = 5, \quad k = 4, \quad \lambda = 1.
$$

It can be verified that a p-dimensional square lattice design with $p < s+1$ is a PBIB design with two associate classes having L_p association scheme with parameters

$$
v = s^2, \ b = ps, \ r = p, \ k = s, \ \lambda_1 = 1, \ \lambda_2 = 0, \ n_1 = p(s - 1),
$$
$$
n_2 = (s - 1)(s - p + 1), \ p_{12}^1 = (s - p + 1)(p - 1), \ p_{12}^2 = p(s - p).
\tag{9.9}
$$

The expressions in the analysis of 2-associate class PBIB designs (see Sec. 8.3) simplify to

$$
\Delta = p(p - 1), \quad c_1 = 1/s, \quad c_2 = -(s - p)/\{s(p - 1)\}
\tag{9.10}
$$

and hence

$$
\hat{\tau}_i = \frac{ps + p - s}{ps(p - 1)} Q_i + \frac{1}{ps(p - 1)} S_1(Q_i),
\tag{9.11}
$$

where Q_i is the ith treatment total minus block means in which ith treatment occurs and $S_1(Q_i)$ is the sum of all Q_i of all treatments which occur together with treatment i. Also,

$$
V(\hat{\tau}_i - \hat{\tau}_j) =
\begin{cases}
2(s + 1)\sigma^2/\{ps\}, & \text{if } (i, j) = 1 \\
2(ps + p - s)\sigma^2/\{ps(p - 1)\}, & \text{if } (i, j) = 2.
\end{cases}
\tag{9.12}
$$

The analysis can be completed by standard methods.

We will give an alternative interpretation for a p-dimensional square lattice. We will identify the s^2 treatments as treatment combinations of a 2-factor experiment consisting of factors a and b each at s levels. Let the rows (columns) correspond to the levels of factor a (factor b). The p-dimensional lattice is then a partially confounded factorial experiment, partially confounding the two main effects, and $p - 2$ sets of pencils of interaction each with $s - 1$ degrees of freedom. The unconfounded interaction sum of squares can be obtained from all replications and the sum of squares of partially confounded interactions is calculated from those replications where they are unconfounded. The sum of squares for treatments adjusted for blocks is the same whether obtained in this manner or the standard method discussed in Chap. 2.

9.3 Simple Triple Lattice

The $v = s^3$ treatments in this case are arranged on an $s \times s \times s$ cube. A replicate of s^2 blocks of size s is obtained by taking the lines parallel to each of the three possible directions. The simple triple lattice can be verified to be a PBIB designs with three associate classes having cubic association scheme and with parameters

$$v = s^3, \ b = 3s^2, \ r = 3, \ k = s, \ \lambda_1 = 1, \ \lambda_2 = \lambda_3 = 0,$$

$$n_1 = 3(s - 1), n_2 = 3(s - 1)^2, \ n_3 = (s - 1)^3. \tag{9.13}$$

It can be verified that in this case

$$\hat{\tau}_i = \frac{2s^2 + 3s + 6}{6s^2} Q_i + \frac{s + 4}{6s^2} S_1(Q_i) + \frac{1}{3s^2} S_2(Q_i) \tag{9.14}$$

and

$$V(\hat{\tau}_i - \hat{\tau}_j) = \begin{cases} 2(s^2 + s + 1)\sigma^2/\{3s^2\}, & \text{if } (i, j) = 1, \\ (2s^2 + 3s + 4)\sigma^2/\{3s^2\}, & \text{if } (i, j) = 2, \\ (2s^2 + 3s + 6)\sigma^2/\{3s^2\}, & \text{if } (i, j) = 3. \end{cases} \tag{9.15}$$

9.4 Rectangular Lattice

The $v = s(s-1)$ symbols are placed in an $s \times s$ array with blank main diagonal. A simple rectangular lattice is formed by having the first replication blocks as the rows of the array with $s - 1$ treatments and the second replication blocks as the columns of the array with $s - 1$ treatments. Nair (1951) showed that a simple rectangular lattice is a PBIB design with four associate classes.

Illustration 9.2 We will construct a simple rectangular lattice for $v = 12$ treatments. We arrange the 12 treatments in a 4×4 array as follows:

$$
\begin{array}{cccc}
- & 1 & 2 & 3 \\
4 & - & 5 & 6 \\
7 & 8 & - & 9 \\
10 & 11 & 12 & -
\end{array}
\tag{9.16}
$$

The design has the following 8 blocks

$$
\begin{array}{llll}
(1, 2, 3); & (4, 5, 6); & (7, 8, 9); & (10, 11, 12); \\
(4, 7, 10); & (1, 8, 11); & (2, 5, 12); & (3, 6, 9).
\end{array}
\tag{9.17}
$$

A triple rectangular lattice is obtained by taking a Latin square with diagonal consisting of all s letters and superimposing on the $s \times s$ array of $s(s-1)$ symbols. The third replication of s blocks is formed by writing the symbols coinciding with each letter of the Latin square.

Illustration 9.2 (cont'd) Let us consider a 4×4 Latin square with main diagonal consisting of symbols A, B, C, D omitted:

$$
\begin{array}{cccc}
- & C & D & B \\
D & - & A & C \\
B & D & - & A \\
A & A & B & -
\end{array}
\tag{9.18}
$$

Superimposing (9.18) onto (9.16), we form the third replication of 4 blocks

$$
(5, 9, 11); \quad (3, 7, 12); \quad (1, 6, 10); \quad (2, 4, 8).
\tag{9.19}
$$

The 12 sets of (9.17) and (9.19) together form a triple rectangular lattice in 12 symbols.

The triple rectangular lattices are PBIB designs when $s = 3$ or 4; but are not PBIB designs when $s \geq 5$.

An r-replicate rectangular lattice in $v = s(s-1)$ treatments can be constructed if $r-1$ MOLS of order s exist. Let $L_1, L_2, \ldots, L_{r-1}$ be a set of MOLS of order s. In L_{r-1} consider the s cells occupied by any symbol and permute the rows and columns of $L_1, L_2, \ldots, L_{r-2}$ such that the s cells are in the diagonal position. In each of $L_1, L_2, \ldots, L_{r-2}$, relabels the symbols such that the diagonal contains the symbols in the natural order. To the 2 replicates of simple rectangular lattice discussed in the beginning add $r - 2$ replicates by superimposing $L_i (i = 1, 2, \ldots, r - 2)$ and forming s blocks with the symbols occurring with each of the s symbols of L_i.

If N is the incidence matrix of a simple rectangular lattice, $N'N$ can be written as

$$N'N = I_2 \otimes (A - B) + J_2 \otimes B, \qquad (9.21)$$

where $\otimes$ is the Kronecker product of matrices, and

$$A = (s - 1)I_s, \quad \text{and} \quad B = J_s - I_s.$$

For more details on lattice designs see Bailey and Speed (1986) and Williams (1977a).

10

Miscellaneous Designs

10.1 α-Designs

In most experiments resolvable block designs with minimum number of replications like 2, 3, or 4 are needed. Lattice designs discussed in Chap. 9, two replicate PBIB designs discussed by Bose and Nair (1962), and resolvable cyclic designs considered by David (1967) are some useful designs. To increase the scope of the designs with high efficiency, Patterson and Williams (1976a) introduced α-designs. These designs can easily be constructed and analyzed using simple computer algorithms.

First let us consider the case, where the number of symbols, v, is a multiple of set size, k, and let $v = ks$ for integral s. The designs are constructed in three steps:

I. Construct a $k \times r$ generating array, α, with elements mod s. This array is called a reduced array if its first row and column has all zeros. Without loss of generality this array can be taken to be a reduced array.

II. Get an intermediate array, α^*, by developing each column of α, mod s as we did in constructing, designs, by the method of differences.

III. In the $k \times rs$ array, α^*, add s to each element of second row, add $2s$ to each element of third row, etc. to get the final design.

Illustration 10.1 We will construct an α-design with $v = 15, k = 3, r = 3$. Here $s = 5$. Consider the generating array, α,

$$
\begin{array}{ccc}
0 & 0 & 0 \\
0 & 1 & 4 \\
0 & 2 & 3
\end{array}
$$

producing the intermediate array α^*,

$$
\begin{array}{ccccccccccccccc}
0 & 1 & 2 & 3 & 4 & 0 & 1 & 2 & 3 & 4 & 0 & 1 & 2 & 3 & 4 \\
0 & 1 & 2 & 3 & 4 & 1 & 2 & 3 & 4 & 0 & 4 & 0 & 1 & 2 & 3 \\
0 & 1 & 2 & 3 & 4 & 2 & 3 & 4 & 0 & 1 & 3 & 4 & 0 & 1 & 2
\end{array}
$$

and giving the final α-design

$$
\begin{array}{ccccccccccccccc}
0 & 1 & 2 & 3 & 4 & 0 & 1 & 2 & 3 & 4 & 0 & 1 & 2 & 3 & 4 \\
5 & 6 & 7 & 8 & 9 & 6 & 7 & 8 & 9 & 5 & 9 & 5 & 6 & 7 & 8 \\
10 & 11 & 12 & 13 & 14 & 12 & 13 & 14 & 10 & 11 & 13 & 14 & 10 & 11 & 12
\end{array}
\tag{10.1}
$$

The columns of (10.1) are the sets of the design with $v = 15 = b, r = 3 = k$. The columns of the design obtained from each column of α is a complete replication of the v symbols.

An α-design is called $\alpha(g_1, g_2, \ldots, g_n)$ design if every pair of symbols occurs together in $g_1, g_2, \ldots,$ or g_n sets. In the design (10.1), every pair of symbols occur together in 0 or 1 sets and hence it is a $\alpha(0, 1)$ design.

Let $C_{\tau|\beta}$ be the C-matrix for estimating treatment effects adjusted for block effects; and $C_{\tau|\beta} = (c_{ij})$. A design is said to be M-S-optimal if $\sum_i c_{ii}$ is maximum and $\sum_{i,j} c_{ij}^2$ is minimum (see Shah and Sinha, 1989). This implies that an $\alpha(0, 1)$ design is preferred over $\alpha(0, 1, 2)$ design and if $\alpha(0, 1, 2)$ has to be used, keep the number of pairs occurring in two sets minimum.

When v is not divisible by k, the derived design with unequal block sizes k_1 and k_2, where $k_2 = k_1 - 1$ can be considered. Let $v = k_1 s_1 + k_2 s_2$, where s_1, s_2 are integers and $k_2 = k_1 - 1$. Construct an α-design in $v + s_2$ symbols $0, 1, \ldots, v - 1,$ $v, \ldots, v + s_2 - 1$ in sets of size k_1 with $s = s_1 + s_2$ sets per replication and delete the symbols $v, v + 1, \ldots, v + s_2 - 1$ from that design.

Efficiency factor, E, of an α-design is taken as r'/r, where σ^2/r' is the average variance of orthonormal contrasts of treatment effects. The maximum efficiency, $E_{\max}$ for a resolvable design in r replications can be calculated from the knowledge of confounded factorial experiments.

In each replication, we may totally lose information on $s - 1$ orthonormal contrasts and estimate them with full information from the other $r - 1$ replications. Hence we get a variance $\sigma^2/(r - 1)$ for $r(s - 1)$ contrasts and a variance σ^2/r for the other $v - r(s - 1) - 1$ contrasts. Hence the average variance

$$
\begin{aligned}
\bar{V} &= \frac{\sigma^2}{v - 1}\left[\frac{r(s - 1)}{r - 1} + \frac{v - r(s - 1) - 1}{r}\right] \\
&= \frac{\sigma^2\{(v - 1)(r - 1) + r(s - 1)\}}{r(r - 1)(v - 1)}.
\end{aligned}
$$

Hence,

$$
E_{\max} = \frac{(v - 1)(r - 1)}{(v - 1)(r - 1) + r(s - 1)}.
\tag{10.2}
$$

The efficiency $E = E_{\max}$ for square lattice designs. Williams (1975a) gave the generating arrays for efficient α-designs with $r = 2, 3, 4$ and set sizes in the range 4 to 16 for $v \leq 100$. For further details see Williams (1975b), Williams, Patterson and John (1976, 1977) and John (1987).

10.2 Trend-Free Designs

Consider equi-replicated, equi-block sized designs, where each symbol occurs at most once in a set.

Blocking and covariance analysis are excellent tools to improve the efficiency of treatment comparisons. In some experimental settings the units within a block may show a trend in one or more directions. In that case the treatment positions may be so chosen that the treatment comparisons are unaffected and we get trend-free block designs. Bradley and Yeh (1980), Chai and Majumdar (1993), Chai and Stufken (1999) Yeh and Bradley (1983), and Yeh, Bradley and Notz (1985) did some pioneering work in this direction.

We consider one directional trend and assume that a polynomial trend of pre-specified degree p common to all blocks exists and is a function of the unit position. Following the notation of Chap. 2, we model the response from jth unit in the ith block, Y_{ij}, as

$$E(Y_{ij}) = \mu + \beta_i + \tau_{d(i,j)} + \sum_{t=1}^{p} \varphi_t(j)\theta_t, \tag{10.3}$$

where $d(i, j)$ is the treatment applied to the jth unit in the ith block, φ_t are orthogonal polynomials of degree t and θ_t are the regression coefficients for trend component φ_t. By stacking the responses as a vector, we can write the complete observational setup (10.3) as

$$E(\mathbf{Y}) = \mathbf{1}_n\mu + (I_b \otimes \mathbf{1}_k)\boldsymbol{\beta} + U\boldsymbol{\tau} + X_\phi\boldsymbol{\theta}, \tag{10.4}$$

where X_ϕ is the design matrix of trend and $\boldsymbol{\theta}$ is the vector of regression coefficients associated with the trend, and the other terms as defined earlier. We have

Theorem 10.1 (*Bradley and Yeh*, 1980) *A necessary and sufficient condition for a block design to be trend-free is*

$$U'X_\phi = 0. \tag{10.5}$$

A simple interpretation of Eq. (10.5) is that the trend component $\varphi_t(j)$ should sum to zero for the set of plots assigned to each treatment.

Table 10.1. ANOVA for a trend-free block design.

Source	df	SS	MS	F	p
Blocks (ign treat)	$b - 1$	SS_B			
Treatments (adj. bl)	$v - 1$	$SS_{Tr|B}$	$MS_{Tr|B}$	$\dfrac{MS_{Tr|B}}{\hat{\sigma}^2}$	p_1
Trend	p	$\sum\limits_{t=1}^{p} W_t^2/b$			
Error	$vr - v - b - p + 1$	by subtraction	$\hat{\sigma}^2$		
Total	$vr - 1$	$\sum\limits_{i,j} Y_{ij}^2 - \dfrac{G^2}{vr}$			

$p_1 = P(F(v - 1, vr - v - b - p + 1) \geq F_{Tr|B}(\text{cal}))$.

The block designs constructed by the method of differences with only 1 symbol attached to the module elements $0, 1, \ldots, m - 1$ and no ∞ symbol used, is clearly a trend-free block design of degree $k - 1$.

The analysis of these designs follows the standard method. Let

$$W_t = \sum_{i=1}^{b} \sum_{j=1}^{k} \varphi_t(j) Y_{ij}, \quad t = 1, 2, \ldots, p.$$

Then $\hat{\theta}_t = W_t/b$. The ANOVA table is given in Table 10.1.

The null hypothesis of equality of treatment effects is rejected when p_1 is less than the significance level.

10.3 Balanced Treatment Incomplete Block Designs

In experiments testing active treatments versus control plays an important role. Dunnett (1955) developed the multiple comparison procedure which is commonly used with a completely randomized design, or randomized block design.

Let o be the control treatment and $1, 2, \ldots, v$ be v active treatments. Occasionally data have to be collected on the $v + 1$ treatments using incomplete blocks of size $k(< v + 1)$ and draw simultaneous inferences on contrasts of the type $\tau_i - \tau_o$ for $i = 1, 2, \ldots, v$. To this end we consider incomplete block designs satisfying:

1. $\text{Var}(\hat{\tau}_i - \hat{\tau}_o) = a^2\sigma^2; i = 1, 2, \ldots, v,$
2. $\text{Cov}(\hat{\tau}_i - \hat{\tau}_o, \hat{\tau}_{i'} - \hat{\tau}_o) = a^2\rho\sigma^2; i, i' = 1, 2, \ldots, v; i \neq i'.$

Clearly 1 and 2 implies

3. $\text{Var}(\hat{\tau}_i - \hat{\tau}_{i'}) = 2a^2(1 - \rho)\sigma^2; i, i' = 1, 2, \ldots, v; i \neq i'.$

In order to satisfy 1, 2, and 3, $C_{\tau|\beta}$ matrix for treatments $0, 1, \ldots, v$, must be of the form

$$C_{\tau|\beta} = \begin{pmatrix} d & e\mathbf{1}'_v \\ e\mathbf{1}_v & fI_v + gJ_v \end{pmatrix}. \tag{10.6}$$

To obtain (10.6), we need to have

$$NN' = \begin{pmatrix} r_0 & \lambda_0\mathbf{1}'_v \\ \lambda_0\mathbf{1}_v & (r_1 - \lambda_1)I_v + \lambda_1 J_v \end{pmatrix}. \tag{10.7}$$

Incomplete block designs whose incidence matrix N satisfies (10.7) are called Balanced Treatment Incomplete Block (BTIB) designs and were introduced by Bechofer and Tamhane (1981) and further considered by Bhaumik (1990), Hedayat and Majumdar (1985), Kim and Stufken (1995), Majumdar (1996), Majumdar and Notz (1983), Notz and Tamhane (1983), and Stufken and Kim (1992).

While Bechofer and Tamhane (1981) considered these designs as maximizing the coverage probability of simultaneous confidence intervals for the control-test treatment contrasts, Stufken (1987) considered them as minimizing the sum of the variances of the estimators. Solorzano and Spurrier (2001) considered designs with more than one control treatment.

Illustration 10.2 Augment the control treatment, 0, to each block of a BIB design with parameters v, b, r, k, λ in treatments $1, 2, \ldots, v$ to get a BTIB design, whose incidence matrix N satisfies

$$NN' = \begin{pmatrix} b & r\mathbf{1}'_v \\ r\mathbf{1}_v & (r - \lambda)I_v + \lambda J_v \end{pmatrix}.$$

Illustration 10.3 Consider the design

$$(0, 1, 3), \quad (0, 2, 3), \quad (0, 3, 4), \quad (0, 0, 1)$$
$$(0, 0, 2), \quad (0, 0, 4), \quad (1, 2, 4)$$

which is a BTIB design with $v = 4, b = 7, k = 3$ with incidence matrix N satisfying

$$NN' = \begin{pmatrix} 15 & 3\mathbf{1}'_4 \\ 3\mathbf{1}_4 & 2I_4 + J_4 \end{pmatrix}.$$

10.4 Nested Block Designs

In some experiments, the blocks may not completely remove the variability and we have to create sub-blocks to conduct the experiment. Preece (1967a) introduced nested balanced incomplete block designs and several authors studied other types of nested designs (see Dey, Das and Banerjee, 1986; Gupta, 1993; Hormel and Robinson, 1975; Jimbo and Kuriki, 1983; Kageyama and Miao, 1997; and Kageyama, Philip and Banerjee, 1995). We will consider the nested balanced incomplete block designs here.

A design in v treatments arranged in b_1 blocks where each block has m sub-blocks, each sub-block of size k_2 is said to be a nested BIB design, if

a) the arrangement of v treatments in $b_2 = b_1 m$ sub-blocks of size k_2 is a BIB design in r replications and λ-parameter λ_2,

b) the arrangement of v treatments in b_1 blocks of size $k_1 = k_2 m$ is a BIB design in r replications and λ-parameter, λ_1.

Clearly

$$b_2 = b_1 m, \quad k_1 = mk_2, \quad vr = b_1 k_1 = b_2 k_2$$

and

$$r(k_1 - 1) = \lambda_1(v - 1); \quad r(k_2 - 1) = \lambda_2(v - 1).$$

Illustration 10.3 We give a nested BIB design with $v = 8$, $b_1 = 14$, $b_2 = 28$, $k_1 = 4$, $k_2 = 2$, $r = 7$, $\lambda_1 = 3$, $\lambda_2 = 1$. We show the blocks by braces and sub-blocks by parentheses. The following is the design:

$$\{(1, 3), (2, 5)\}; \quad \{(2, 3), (5, 7)\}; \quad \{(2, 4), (3, 6)\};$$
$$\{(3, 4), (6, 7)\}; \quad \{(3, 5), (4, 0)\}; \quad \{(4, 5), (0, 7)\};$$
$$\{(4, 6), (5, 1)\}; \quad \{(5, 6), (1, 7)\}; \quad \{(5, 0), (6, 2)\};$$
$$\{(6, 0), (2, 7)\}; \quad \{(6, 1), (0, 3)\}; \quad \{(0, 1), (3, 7)\};$$
$$\{(0, 2), (1, 4)\}; \quad \{(1, 2), (4, 7)\}.$$

If Y_{ijt} is the response from the tth unit of jth sub-block of the ith block, we assume the model,

$$Y_{ijt} = \mu + \beta_i + \eta_{ij} + \tau_{d(ij,t)} + e_{ijt}, \tag{10.8}$$

where μ is the general mean, β_i is the ith block effect, η_{ij} is the jth sub-block effect nested in the ith block, $\tau_{d(ij,t)}$ is the treatment effect of the treatment applied to the tth unit in the jth sub-block of the ith block, and e_{ijt} are random errors assumed to be independently and identically distributed as $N(0, \sigma^2)$. The information matrix for estimating treatment effects adjusted for block and sub-block effects, $C_{\tau|\beta,\eta}$, can easily be verified to be

$$C_{\tau|\beta,\eta} = rI_v - \frac{1}{k_1}N_1N_1' - \frac{1}{k_2}N_2N_2' + \frac{r}{v}J_v, \tag{10.9}$$

where N_1 (N_2) are incidence matrices of treatments and blocks (sub-blocks).

Suppose we want to arrange a bridge tournament with $4t + 1$ players such that

1. There are $4t + 1$ rounds, where the ith player will not participate in the ith round.
2. In each round there are t tables with 4 players at each table.
3. Each player plays in each of the 4 positions N, E, S, W at each table exactly once.
4. Each player partners with each other player in exactly one round and plays against every player in exactly 2 rounds.

The solution is based on nested BIB design with parameters

$$v = 4t + 1, \quad b_1 = t(4t + 1), \quad k_2 = 2,$$
$$\lambda_2 = 1, \quad \lambda_1 = 3, \quad m = 2, \quad k_1 = 4, \quad r = 4t.$$

Let $4t + 1$ be a prime or prime power and x be a primitive root of GF($4t + 1$). The t initial sets

$$\{(x^i, x^{2t+i}), (x^{t+i}, x^{3t+i})\}, \quad i = 0, 1, \ldots, t - 1$$

when developed mod ($4t + 1$) gives the solution. We interpret the set $\{$(N, S), (E, W)$\}$ as the players in the 4 positions.

With 5 players, $x = 2$ and the solution is

$$\{(1, 4), (2, 3)\},$$
$$\{(2, 0), (3, 4)\},$$
$$\{(3, 1), (4, 0)\},$$
$$\{(4, 2), (0, 1)\},$$
$$\{(0, 3), (1, 2)\}.$$

A pitch tournament (see Finizio and Lewis, 1999) for $8t+1$ players is a schedule of $8t + 1$ rounds of t games such that a team of 4 players plays against a team of 4 players satisfying

(i) the ith player will not play in the ith round;
(ii) each player partners every other player exacty 3 times;
(iii) each player opposes every other player exactly 4 times.

The arrangement of the games and rounds is based on a nested BIB design with parameters

$$v = 8t + 1, \quad b_1 = t(8t + 1), \quad k_2 = 4,$$
$$\lambda_2 = 3, \quad \lambda_1 = 7, \quad m = 2, \quad k_1 = 8, \quad r = 8t.$$

When $8t + 1$ is a prime or prime power, a solution of this form of pitch tournament are the games corresponding to the sets developed $\mod(8t + 1)$ from the sets, corresponding to the initial rounds

$$\left\{ \left(x^i, x^{i+2t}, x^{i+4t}, x^{i+6t} \right), \left(x^{i+t}, x^{i+3t}, x^{i+5t}, x^{i+7t} \right) \right\}; \quad i = 0, 1, \ldots, t - 1,$$

where x is a primitive root of GF$(8t + 1)$.

Nested BIB designs are also used in triallel-cross experiments (see Bhar and Dey, 2004).

We will now give another type of nesting in block designs. Suppose the experimenter wants to test v treatments and no suitable block design exists. However, a design exists with the parameters v_1, b_1, r_1, k_1, where $v_1 < v < 2v_1$. Let $d = 2v_1 - v$. Divide the v treatments into 3 sets T_1, T_2 and T_3 with cardinalities d, $v_1 - d$ and $v_1 - d$ respectively. Form a design D_1 with treatments $T_1 \cup T_2$, and a design D_2 with treatments $T_1 \cup T_3$ and juxtapose them to form $2b_1$ blocks. Let $C_{\tau|\beta}^{(i)}$ be the C-matrix for estimating treatment effects from design D_i, and let

$$C_{\tau|\beta}^{(i)} = \begin{pmatrix} C_{11}^{(i)} & C_{12}^{(i)} \\ C_{21}^{(i)} & C_{22}^{(i)} \end{pmatrix},$$

where $C_{11}^{(i)}$ is the $d \times d$ matrix corresponding to the treatments of T_1. The C-matrix of all v treatments adjusting for $2b_1$ block effects can then be verified as

$$C_{\tau|\beta} = \begin{pmatrix} C_{11}^{(1)} + C_{11}^{(2)} & C_{12}^{(1)} & C_{12}^{(2)} \\ C_{21}^{(1)} & C_{22}^{(1)} & 0 \\ C_{21}^{(2)} & 0 & C_{22}^{(2)} \end{pmatrix}. \tag{10.10}$$

10.5 Nearest Neighbor Designs

In serology experiments circular blocks arise as the treatment applied in a particular position may be affected by the treatments applied in the two adjacent positions. In such cases, we need to balance the treatments in adjacent positions. Given t sets in v symbols $0, 1, \ldots, v - 1$, with the ith set $S_i = \{\theta_{i0}, \theta_{i1}, \ldots, \theta_{i,k-1}\}, i = 1, 2, \ldots, t$; the design $S_{i\theta} = S_i + \theta$ for $\theta = 0, 1, \ldots, v - 1; i = 1, 2, \ldots, t$; is called nearest neighbor balance design if the differences $\theta_{ij} - \theta_{i,j+1}, \theta_{ij} - \theta_{i,j-1}$

mod v contains all the $v - 1$ symbols $1, 2, \ldots, v - 1$ equal number of times, for $j = 0, 1, \ldots, k - 1; i = 1, 2, \ldots, t$, with the understanding that $\theta_{ik} = \theta_{i0}$.

Illustration 10.4 The BIB design

$$(0, 1, 3); \quad (1, 2, 4); \quad (2, 3, 5); \quad (3, 4, 6); \quad (4, 5, 0); \quad (5, 6, 1); \quad (6, 0, 2)$$

is nearest neighbor balanced with $t = 1$. The indicated differences contain $1, 2, \ldots, 6$ exactly once.

Illustration 10.5 The design

$$(0, 1, 3, 1); \quad (1, 2, 4, 2); \quad (2, 3, 0, 3); \quad (3, 4, 1, 4); \quad (4, 0, 2, 0)$$

is also nearest neighbor balanced.

In a nearest neighbor balanced design, the sets need not contain all distinct symbols.

Different models can be used to analyze data coming from such designs and the interested reader is referred to Besag and Kemptom (1986). Also, see Chai and Majundar (2000) for other optimality results.

10.6 Augmented Block Designs

When new treatments are introduced, there might not be enough material to replicate the new treatments. In such cases the new treatments will be combined with standard (check) treatments using a single replication of each of the new treatments. Such designs are called augmented designs (see Federer and Raghavarao, 1975).

The statistical analysis for a block design in which v check varieties are used replicating ith variety r_i times and v^* new varieties replicated once can be carried in two ways.

(a) As a general block design using $v + v^*$ treatments and drawing inferences on the contrasts of interest.
(b) Analyze the check varieties data only getting estimates of block effects and error variance. The response on the new varieties are adjusted for block effects and inferences are drawn.

Federer and Raghavarao (1975) showed that the two methods of analyses are the same. The randomization procedure (see Federer, 1961) is to follow the standard procedure for check varieties and all new varieties are assigned randomly to the remaining units. Assume that we want to use 7 check varieties in a linked block design with 7 blocks and use 10 new varieties. Let $A, B, \ldots, G$ be the check varieties. The following is the plan.

Blocks	Block Contents				
1	A	B	$\underline{1}$	D	
2	$\underline{2}$	B	E	C	
3	D	C	$\underline{3}$	$\underline{4}$	F
4	E	D	G		
5	E	$\underline{5}$	$\underline{6}$	F	A
6	7	F	G	B	$\underline{8}$
7	$\underline{9}$	G	$\underline{10}$	A	C

By analyzing the LB design in check varieties, contrasts of $\beta_1, \ldots, \beta_7$ can be estimated. Using Y_{ij} as the response of the jth unit in the ith block, we have

$$(\hat{\tau}_1 - \hat{\tau}_2) = Y_{13} - \hat{\beta}_1 - (Y_{21} - \hat{\beta}_2) = Y_{13} - Y_{21} - (\hat{\beta}_1 - \hat{\beta}_2),$$
$$(\hat{\tau}_3 - \hat{\tau}_4) = Y_{33} - Y_{34}.$$

Clearly

$$\mathrm{Var}(\hat{\tau}_1 - \hat{\tau}_2) = 2\sigma^2 + \mathrm{Var}(\hat{\beta}_1 - \hat{\beta}_2) = \left(2 + \frac{6}{7}\right)\sigma^2$$

and $\mathrm{Var}(\hat{\tau}_3 - \hat{\tau}_4) = 2\sigma^2$. σ^2 is estimated from the check varieties data.

10.7 Computer Aided Block Designs

There are many computer modules that provide optimal block designs. In this section we describe the GENDEX module of D.O.E. toolkit of Nguyen to construct optimal equi-block sized, equi-replicated designs. Consider an incomplete block design with parameters $v, b, r, k(< v)$. The efficiency of the design discussed earlier can be written as

$$E = (v - 1) \left/ \sum_{i=1}^{v-1} e_i^{-1} \right. \tag{10.11}$$

for a connected design, where e_i are nonzero eigenvalues of $r^{-1} C_{\tau|\beta}$. Maximizing E is the same as minimizing $\sum_{i=1}^{v-1} e_i^{-1}$, and noting that

$$r^{-1} C_{\tau|\beta} = I - (rk)^{-1} NN', \tag{10.12}$$

 Block Designs: Analysis, Combinatorics and Applications

we have

$$\sum e_i^{-1} = \text{constant} + \sum_{i=2}^{\infty} (rk)^{-i} \text{tr}(NN')^i, \tag{10.13}$$

where N is the incidence matrix of the block design. Instead of minimizing $\sum e_i^{-1}$, the strategy is to minimize the first two terms of (10.13). The first stage is to minimize the first term of the right-hand side of (10.13) given by

$$f_2 = \sum_{i=1}^{v-1} \sum_{i'=i+1}^{v} a_{ii'}^2, \tag{10.14}$$

where $NN' = A = (a_{ii'})$. Designs that minimize f_2 are (M-S)-optimal designs. The aim here is to get a BIB design or a regular graph design (RGD) with only two $a_{ii'}$ terms for $i \neq i'$ differing by 1. If such a design is found, go to stage 2, and minimize

$$f_3 = \sum_{i=1}^{v-2} \sum_{i'=i+1}^{v-1} \sum_{i''=i'+1}^{v} a_{ii'} a_{ii''} a_{i'i''} \tag{10.15}$$

while keeping the RGD status unaltered. Lower bounds on f_2 and f_3 are known. The algorithm used is to form a design randomly with v, r, k parameters and swap the treatments between every pair of blocks so that f_2 and f_3 are minimized, to reach the lower bound or as close as possible to the lower bound.

Nguyen (1994) claims that his program found 45 of the 63 known BIBD's in the range $v \leq 100, r, k \leq 10$ and the remaining 18 are within 0.5% of optimality.

His program also found the solution of the BIB design with parameters

$$v = 15, \quad b = 42, \quad r = 14, \quad k = 5, \quad \lambda = 4$$

which was listed as unsolved in Raghavarao (1971). A solution of this BIB design by the method of differences is also indicated in Table 4.4.

10.8 Design for Identifying Differentially Expressed Genes

Microarray experiments are widely used these days to determine the expression levels of thousands of genes. The purpose of those experiments is to identify the genes with different levels for normal people and people with an abnormal condition. When multiple observations are available on a gene under the two conditions, a two-sample t-test or a permutation test can be performed to test whether the two expression levels are different. However, when the abnormal condition is rare, multiple observations may not be available to perform a two-sample t-test.

Surprisingly, we can use some designs to create multiple observations and test that a gene is differently expressed under the two conditions.

Consider an affine resolvable BIB design with parameters

$$
\begin{aligned}
v &= n^2[(n-1)t + 1], \quad b = n(n^2 t + n + 1), \\
r &= n^2 t + n + 1, \qquad k = n[(n-1)t + 1], \\
\lambda &= nt + 1
\end{aligned}
$$

with n blocks per replication. Let S_{ij} be the jth set in the ith replication; $j = 1, 2, \ldots, n$; $i = 1, 2, \ldots, r$. Without loss of generality assume that symbol 1 occurs in the set S_{i1} for every i. We use the $2r$ sets S_{i1}, S_{i2} for $i = 1, 2, \ldots, r$ to identify the differentially expressed genes.

Assume that we have one gene to be tested whether it has different expression levels under the two conditions and let us label it as gene 1. Along with this gene we take $v - 1$ genes, which are known to have the same expression levels under the two conditions. Let x_i and y_i be appropriately normalized expression levels for gene i under the normal and abnormal conditions, respectively. If x_i and y_i are far apart, then gene i is differentially expressed; otherwise, it is similarly expressed. For this purpose, we define

$$
u_i = \log_2 \left(\frac{y_i}{x_i} \right).
$$

We assume $u_i \sim \mathrm{IN}(0, \sigma^2)$ for $i = 2, 3, \ldots, v$. u_1 will be independently distributed from u_i ($i = 2, 3, \ldots, v$), with mean θ. The variance of u_1 may or may not be equal to σ^2, and we discuss these two cases separately.

Case 1. $u_1 \sim N(\theta, \sigma^2)$.

We want to test the null hypothesis $H_0\colon \theta = 0$, versus the alternative $H_A\colon \theta \neq 0$. The rejection of H_0 implies that gene 1 is differentially expressed under the two conditions.

We transform u_i ($i = 1, 2, \ldots, v$) into r observations w_j ($j = 1, 2, \ldots, r$) as follows:

$$
w_j = \sum_{a \in S_{j1}} u_a - \sum_{b \in S_{j2}} u_b.
$$

Clearly $w_j \sim \mathrm{IN}(\theta, 2k\sigma^2)$. Independence follows because S_{j1} and S_{j2} are two sets of the same replicate in an affine resolvable BIB design. We thus have a random sample $\{w_1, w_2, \ldots, w_r\}$ from a normal population with mean θ, and the hypothesis

$H_0: \theta = 0$ can be tested by a one-sample t-test. Let

$$\bar{w} = \frac{1}{r}\sum_j w_j, \quad s_w = \sqrt{\frac{\sum_j (w_j - \bar{w})^2}{r-1}}.$$

The test statistic is

$$t = \frac{\bar{w}\sqrt{r}}{s_w} \tag{10.16}$$

and the p-value for testing the hypothesis is

$$p\text{-value} = 2P(t(r-1) > |t_{\text{cal}}|),$$

where t_{cal}, is the calculated test statistic (10.16).

Usually several genes may have to be tested for differential expression, and we need to consider the multiplicity and adjust the p-value to determine the significance of the tested gene.

Case 2. $u_1 \sim N(\theta, (1+\delta\theta^2)\sigma^2)$

The genes with large mean for u_i may have larger variance than the unexpressed genes and we assume $\text{var}(u_1) = (1+\delta\theta^2)\sigma^2$ for a known constant δ. For simplicity we take $\delta = 1$. We need an estimate of σ^2 from all genes known to have similar expression levels under both conditions and let us take this as $\hat{\sigma}^2$. The quantity considered in case 1 will have the form

$$\bar{w} = u_1 + \sum_{i=2}^{v} d_i u_i$$

for suitable rational d_i's. Also,

$$\text{Var}(\bar{w}) = \left\{ 1 + \delta\theta^2 + \sum_{i=2}^{v} d_i^2 \right\} \sigma^2.$$

Thus the null hypothesis $H_0: \theta = \theta_0$ will not be rejected against a two-sided alternative, if

$$\frac{(\bar{w} - \theta_0)^2}{\left(1 + \delta\theta_0 + \sum_{i=2}^{v} d_i^2\right)\hat{\sigma}^2} \leq \chi^2_{1-\alpha}(1). \tag{10.17}$$

Equation (10.17) will simplify to

$$A\theta_0^2 + B\theta_0 + C \leq 0.$$

Let $D = B^2 - 4AC$ be the discriminant of the quadratic equation and let us assume that D is positive. Let $\theta_1 = (-B - \sqrt{D})/2A$, and $\theta_2 = (-B + \sqrt{D})/2A$. When $A > 0$, we have $\theta_1 < \theta_2$ and $\theta_1 \leq \theta_0 \leq \theta_2$; and when $A < 0$, we have $\theta_2 < \theta_1$ and $\theta_0 \leq \theta_2$ or $\theta_0 \geq \theta_1$.

We thus conclude that gene 1 has differential expression, if

1. $A > 0$, and $0 \notin [\theta_1, \theta_2]$; or
2. $A < 0$, and $0 \in [\theta_2, \theta_1]$.

While testing several genes, α of Eq. (10.17) can be modified to account for the multiplicity problem.

In microarray experiments the data on $\log_2(y_i/x_i)$ can be arranged in increasing order. The middle part can be assumed to correspond to the unexpressed genes and each gene expression from the top and bottom can be tested by the methods described here. The method described above can be used with multiple arrays also by taking x_i and y_i to be the mean expression levels.

The affine resolvable BIB design with two sets per replication without repeated blocks necessarily has the parameters

$$v = 4t, \quad b = 2(4t - 1), \quad r = 4t - 1, \quad k = 2t, \quad \lambda = 2t - 1$$

and using this design, we will have large degrees of freedom for the t-test of case 1 and we take fewer unexpressed genes in the study. This design is closely related to Hadamard matrix, H_{4t}. Ding and Raghavarao (2005) used Hadamard matrix to identify the expressed genes and illustrated their method with APOAI data. They considered the middle 80% of $\log_2(y_i/x_i)$ to consist of unexpressed genes and tested each of the top 10% and bottom 10% for different expression levels. They used 7, 15, 31 and 63 unexpressed genes to test the doubtful genes and examined the coefficient of variation on the number of determined differentially expressed genes. Empirically they recommend the use of 31 unexpressed genes with a coefficient of variation of 0.1 to test each doubtful gene in a microarray experiment. For a brief review of work on this topic, the interested reader is referred to Dudoit, Yang, Callow, and Speed (2002).

10.9 Symmetrical Factorial Experiments with Correlated Observations in Blocks

Consider an s^n factorial experiment in n factors with each factor at s levels. We assume the levels to be equally spaced and denote them by $a_1, a_2, \ldots, a_s$, such that $\sum a_i = 0$. We use the full factorial treatment combinations and assume a main effects model. We want to arrange the s^n treatment combinations in b blocks of sizes $k_1, k_2, \ldots, k_b$ $\left(\sum k_i = s^n\right)$. We assume that a pair of observations in the same block is positively correlated and observations in different blocks are uncorrelated. Let x_{ijh} be the level of hth factor in the jth observation in the ith block for $h = 1, 2, \ldots, n;\ j = 1, 2, \ldots, k_i;\ i = 1, 2, \ldots, b$. Using straightforward algebra, Sethuraman, Raghavarao and Sinha (2005) showed that the D-optimal design for estimating main effects with $k_i = d_i$s is the one with $\sum_j x_{ijh} = 0$, for every i and h.

Specializing to 2^n experiments, we can construct D-optimal block designs with even block sizes. For any treatment combination (run), by interchanging the levels of all factors, we get a fold-over run. There are 2^{n-1} fold-over pairs of runs in a 2^n factorial experiment. Let 1 and -1 be the 2 levels of each factor and $(a_1, a_2, \ldots, a_n)$ be a run with n factors at levels $a_1, a_2, \ldots, a_n$ respectively. For a 2^4 experiment, the eight fold-over pairs of runs are:

$$\begin{aligned}
&\{(-1, -1, -1, -1);\ (1, 1, 1, 1)\}, \\
&\{(-1, -1, -1, 1);\ (1, 1, 1, -1)\}, \\
&\{(-1, -1, 1, -1);\ (1, 1, -1, 1)\}, \\
&\{(-1, -1, 1, 1);\ (1, 1, -1, -1)\}, \\
&\{(-1, 1, -1, -1);\ (1, -1, 1, 1)\}, \\
&\{(-1, 1, -1, 1);\ (1, -1, 1, -1)\}, \\
&\{(-1, 1, 1, -1);\ (1, -1, -1, 1)\}, \\
&\{(-1, 1, 1, 1);\ (1, -1, -1, -1)\}.
\end{aligned}$$

$$(10.18)$$

We can form a D-optimal design with even block sizes using the fold-over pairs in blocks. We can form a block design in 4 blocks of size 4 by putting pairs of fold-over pairs as follows:

$$\begin{aligned}
&\{(-1, -1, -1, -1);\ (1, 1, 1, 1);\ (-1, -1, -1, 1);\ (1, 1, 1, -1)\}, \\
&\{(-1, -1, 1, -1);\ (1, 1, -1, 1);\ (-1, -1, 1, 1);\ (1, 1, -1, -1)\}, \\
&\{(-1, 1, -1, -1);\ (1, -1, 1, 1);\ (-1, 1, -1, 1);\ (1, -1, 1, -1)\}, \\
&\{(-1, 1, 1, -1);\ (1, -1, -1, 1);\ (-1, 1, 1, 1);\ (1, -1, -1, -1)\}.
\end{aligned}$$

Sethuraman, Raghavarao and Sinha (2005) further showed that for a 2^n experiment with even block sizes except for a pair of blocks of odd sizes the design with $\sum_j x_{ijh} = 0$, for even sized blocks and $\sum_j x_{ijh} = \pm 1$ for odd sized blocks is D-optimal. This implies that the blocks will be made up of fold-over pairs except that one pair will be split and the runs are separately augmented to the two odd sized blocks. We want to construct a D-optimal 2^4 experiment in 4 blocks of sizes $k_1 = 3, k_2 = 5, k_3 = 4, k_4 = 4$. We form blocks as before using one fold-over pair for block 1, and two-fold over pairs for each of blocks 2, 3, and 4. We then use the first run of last pair of (10.18) to block 1 and second run of last pair to block 2. The design so constructed is

$$\{(-1, -1, -1, -1); (1, 1, 1, 1); (-1, 1, 1, 1)\},$$
$$\{(-1, -1, -1, 1); (1, 1, 1, -1); (-1, -1, 1, -1);$$
$$(1, 1, -1, -1); (1, -1, -1, -1)\},$$
$$\{(-1, -1, 1, 1); (1, 1, -1, -1); (-1, 1, -1, -1); (1, -1, 1, 1)\},$$
$$\{(-1, 1, -1, 1); (1, -1, 1, -1); (-1, 1, 1, -1); (1, -1, -1, 1)\}.$$

For other results in this connection see Atkins and Cheng (1995), Cheng and Steinberg (1991), Goos (2002), Russell and Eccleston (1987a,b).

References and Selected Bibliography

Adhikary, B. (1966). Some types of PBIB association schemes, *Calcutta Statist. Assoc. Bull.* **15**, 47–74.

Adhikary, B. (1967). A new type of higher associate cyclical association scheme, *Calcutta Statist. Assoc. Bull.* **16**, 40–44.

Adhikary, B. (1972a). Method of group differences for the construction of three associate cyclic designs, *J. Indian Soc. Agric. Statist.* **24**, 1–18.

Adhikary, B. (1972b). A note on restricted Kronecker product method of constructing statistical designs, *Calcutta Statist. Assoc. Bull.* **21**, 193–196.

Adhikary, B. (1973). On generalized group divisible designs, *Calcutta Statist. Assoc. Bull.* **22**, 75–88.

Agarwal, G.G. and Kumar, S. (1984). On a class of variance balanced designs associated with group divisible designs, *Calcutta Statist. Assoc. Bull.* **33**, 178–190.

Aggarwal, H.C. (1977). A note on the construction of partially balanced ternary designs, *J. Indian Soc. Agric. Statist.* **29**, 92–94.

Aggarwal, K.R. (1972). A note on construction of triangular PBIB design with parameters $v = 21, b = 35, r = 10, k = 6, \lambda_1 = 2, \lambda_2 = 3$, *Ann. Math. Statist.* **43**, 371.

Aggarwal, K.R. (1973). Analysis of $ML_i(s)$ designs, *J. Indian Soc. Agric. Statist.* **25**, 215–218.

Aggarwal, K.R. (1974). Analysis of $L_i(s)$ and triangular designs, *J. Indian Soc. Agric. Statist.* **26**, 3–13.

Aggarwal, K.R. (1975). Modified Latin square type PBIB designs, *J. Indian Soc. Agric. Statist.* **27**, 49–54.

Aggarwal, K.R. (1977). Block structure of certain series of EGD and hyphercubic designs, *J. Indian Soc. Agric. Statist.* **29**, 19–23.

Aggarwal, K.R. and Raghavarao, D. (1972). Residue classes PBIB designs, *Calcutta Statist. Assoc. Bull.* **21**, 63–69.

Agrawal, H.L. and Boob, B.S. (1976). A note on method of construction of affine resolvable balanced incomplete block designs, *Ann. Statist.* **4**, 954–955.

Alltop, W.O. (1969). An infinite class of 4-designs, *J. Comb. Th.* **6**, 320–322.

Alltop, W.O. (1971). Some 3-designs and a 4-design, *J. Comb. Th.* **11A**, 190–195.

Alltop, W.O. (1972). An infinite class of 5-designs, *J. Comb. Th.* **12A**, 390–395.

Alltop, W.O. (1975). Extending t-designs, *J. Comb. Th.* **18A**, 177–186.

Anderson, I. (1997). *Combinatorial Designs and Tournaments*, Oxford University Press, London.

Archbold, J.W. and Johnson, N.L. (1956). A method of constructing partially balanced incomplete block designs, *Ann. Math. Statist.* **27**, 624–632.

Arnab, R. and Roy, D. (1990). On use of symmetrical balanced incomplete block design in construction of sampling design realizing pre-assigned sets of inclusion probabilities of first two orders, *Comm. Statist.* **19A**, 3223–3232.

Aschbacher, M. (1971). On collineation groups of symmetrical block designs, *J. Comb. Th.* **3**, 272–281.

Assmus, E.F. (Jr.) and Mattson, H.F. (Jr.) (1969). New 5-designs, *J. Comb. Th.* **6**, 122–151.

Atkins, J.E. and Cheng, C.-S. (1995). Optimal regression designs in the presence of random block effect, *J. Statist. Plan. Inf.* **77**, 321–335.

Atkinson, A. and Donev, A. (1992). *Optimum Experimental Designs*, Oxford University Press, Oxford.

Avadhani, M.S. and Sukhatme, B.V. (1973). Controlled sampling with equal probabilities and without replacement, *Int. Statist. Rev.* **41**, 175–183.

Bagchi, B. and Bagchi, S. (2001). Optimality of partial geometric designs, *Ann. Statist.* **29**, 577–594.

Bagchi, S. (1988). A class of non-binary unequally replicated E-optimal designs, *Metrika* **35**, 1–12.

Baker, R.D. (1976). Self-orthogonal Latin squares and BIBD's, *Proc. 7th S.E. Conf. Comb., Graph Th. Computing*, 101–114.

Baker, R.D. (1983). Resolvable BIBD and SOLS, *Discr. Math.* **44**, 13–29.

Balakrishnan, R. (1971). Multiplier groups of difference sets, *J. Comb. Th.* **10A**, 133–139.

Bailey, R.A. and Speed, T.P. (1986). Rectangular lattice designs: Efficiency factors and analysis, *Ann. Statist.* **14**, 874–895.

Bartmans, A. and Shrikhande, M.S. (1981). Tight subdesigns of symmetric designs, *Ars Comb.* **12**, 303–309.

Baumert, L.D. (1971). *Cyclic Difference Sets*, Lecture Notes in Mathematics 182, Springer-Verlag, Berlin.

Bechhofer, R.E. and Tamhane, A.C. (1981). Incomplete block designs for comparing treatments with a control: General theory, *Technometrics* **23**, 45–57.

Becker, H. and Haemers, W. (1980). 2-designs having an intersection number $k - n$, *J. Comb. Th.* **28A**, 64–81.

Benjamini, Y. and Hochberg, Y. (1995). Controlling the false discovery rate: A practical and powerful approach to multiple testing, *J. Roy. Statist. Soc.* **57B**, 289–300.

Besag, J. and Kempton, R. (1986). Statistical analysis of field experiments using neighboring plots, *Biometrics* **42**, 231–251.

Beth, T., Jungnickel, D. and Lenz, H. (1999). *Design Theory*, Vols. I & II, Cambridge University Press, Cambridge.

Bhagwandas and Sastry, D.V.S. (1977). Some constructions of balanced incomplete block designs, *J. Indian Statist. Assoc.* **15**, 91–95.

Bhar, L. and Dey, A. (2004). Triallel-cross block designs that are robust against missing observations, *Utilitas Math.* **65**, 231–242.

Bhat, V.N. (1972a). Non-isomorphic solutions of some balanced incomplete block designs. II, *J. Comb. Th.* **12A**, 217–224.

Bhat, V.N. (1972b). Non-isomorphic solutions of some balanced incomplete block designs. III, *J. Comb. Th.* **12A**, 225–252.

Bhattacharya, K.N. (1944). A new balanced incomplete block design, *Sci. Cult.* **9**, 508.

Bhaumik, D.K. (1990). Optimal incomplete block designs for comparing treatments with a control under the nearest neighbor correlation model, *Utilitas Math.* **38**, 15–25.

Bhaumik, D.K. (1995). Optimality in the competing effects model. *Sankhyā* **57B**, 48–56.

Blackwedder, W.C. (1969). On constructing balanced incomplete block designs from association matrices with special reference to association schemes of two and three classes, *J. Comb. Th.* **17**, 15– 36.

Bondar, J.V. (1983). Universal optimality of experimental designs: Definitions and a criterion, *Canad. J. Statist.* **11**, 325–331.

Boob, B.S. (1976). On a method of construction of symmetrical BIB design with symmetrical incidence matrix, *Calcutta Statist. Assoc. Bull.* **25**, 175–178.

Bose, R.C. (1939). On the construction of balanced incomplete block designs, *Ann. Eugen.* **9**, 353–399.

Bose, R.C. (1942). A note on the resolvability of balanced incomplete designs, *Sankhyā* **6**, 105–110.

Bose, R.C. (1949). A note on Fisher's inequality for balanced incomplete block designs, *Ann. Math. Statist.* **20**, 619–620.

Bose, R.C. (1963). Strongly regular graphs, partial geometries and partially balanced designs, *Pacific J. Math.* **13**, 389–419.

Bose, R.C. (1977). Symmetric group divisible designs with the dual property, *J. Statist. Plan. Inf.* **1**, 87–101.

Bose, R.C. (1979). Combinatorial problems of experimental designs I. Incomplete block designs, *Proc. Symposia in Pure Mathematics* **34**, 47–68.

Bose, R.C. and Connor, W.S. (1952). Combinatorial properties of group divisible incomplete block designs, *Ann. Math. Statist.* **23**, 367–383.

Bose, R.C. and Laskar, R. (1967). A characterization of tetrahedral graphs, *J. Comb. Th.* **2**, 366–385.

Bose, R.C. and Mesner, D.M. (1959). On linear associative algebras corresponding to association schemes of partially balanced designs, *Ann. Math. Statist.* **30**, 21–38.

Bose, R.C. and Nair, K.R. (1939). Partially balanced incomplete block designs, *Sankhyā* **4**, 337–372.

Bose, R.C. and Nair, K.R. (1962). Resolvable incomplete block designs with two replications, *Sankhyā* **24A**, 9–24.

Bose, R.C. and Shimamoto, T. (1952). Classification and analysis of partially balanced designs with two associate classes, *J. Amer. Statist. Assoc.* **47**, 151–184.

Bose, R.C. and Shrikhande, M.S. (1979). On a class of partially balanced incomplete block designs, *J. Statist. Plan. Inf.* **3**, 91–99.

Bose, R.C. and Shrikhande, S.S. (1960). On the composition of balanced incomplete block designs, *Canad. J. Math.* **12**, 177–188.

Bose, R.C. and Shrikhande, S.S. (1976). Baer subdesign of symmetric balanced incomplete block designs, *Essays in Prob. Statist.* (eds. S. Ikeda *et al.*), 1–16.

Bose, R.C., Bridges, W.G. and Shrikhande, M.S. (1976). A characterization of partial geometric designs, *Disc. Math.* **16**, 1–7.

Bose, R.C., Bridges, W.G. and Shrikhande, M.S. (1978). Partial geometric designs and two-class partially balanced designs, *Discr. Math.* **21**, 97–101.

Bose, R.C., Shrikhande, S.S. and Singhi, N.M. (1976). Edge regular multi-graphs and partial geometric designs with an application to the embedding of quasi-residual designs, *Proc. Int. Colloquium Comb. Th.* (Rome, September 1973) **1**, 49–81.

Box, G.E.P. and Behnken, D.W. (1960). Some new three level designs for the study of quantitative variables, *Technometrics* **2**, 455–475.

Box, G.E.P. and Cox, D.R. (1964). An analysis of transformation, *J. Roy. Statist. Soc.* **26B**, 211–246.

Box, G.E.P. and Hunter, J.S. (1957). Multifactor experimental designs for exploring response surfaces, *Ann. Math. Statist.* **28**, 195–241.

Bradley, R.A. and Yeh, C.-M. (1980). Trend-free block designs: Theory, *Ann. Statist.* **8**, 883–893.

Brindley, D.A. and Brandley, R.A. (1985). Some new results on grubbs estimators, *J. Amer. Statist. Assoc.* **80**, 711–714.

Brouwer, A.E., Schrijver, A. and Hanani, H. (1978). Group divisible designs with block size four, *Discr. Math.* **20**, 1–10.

Brown, R.B. (1975). A new non-extensible (16, 24, 9, 6, 3)-design, *J. Comb. Th.* **19A**, 115–116.

Bueno Filho, J.S. Des. and Gilmour, S.G. (2003). Planning incomplete block experiments with treatments are genetically related, *Biometrics* **59**, 375–381.

Bush, K.A. (1977a). Block structure properties in combinatorial problems, *Ars Comb.* **4**, 37–41.

Bush, K.A. (1977b). An infinite class of semi-regular designs, *Ars Comb.* **3**, 293–296.

Bush, K.A. (1977c). Bounds for the number of common symbols in balanced and certain partially balanced designs, *J. Comb. Th.* **23A**, 46–51.

Bush, K.A. (1979). Families of regular designs, *Ars Comb.* **7**, 195–204.

Bush, K.A., Federer, W.T., Pesotan, H. and Raghavarao, D. (1984). New combinatorial designs and their applications to group testing, *J. Statist. Plan. Inf.* **10**, 335–343.

Calinski, T. (1971). On some desirable patterns in block designs, *Biometrics* **27**, 275–292.

Calinski, T. and Ceranka, B. (1974). Supplemented block designs, *Biom. Z.* **16**, 299–305.

Calinski, T. and Kageyama, S. (2000). *Block Designs: A Randomization Approach: Analysis*, Vol. I, Springer Lecture Notes in Statistics 150, New York.

Calinski, T. and Kageyama, S. (2003). *Block Designs: A Randomization Approach: Design*, Vol. II, Springer Lecture Notes in Statistics 170, New York.

Calvin, J.A. (1986). A new class of variance balanced designs, *J. Statist. Plan. Inf.* **14**, 251–254.

Calvin, J.A. and Sinha, K. (1989). A method of constructing variance balanced designs, *J. Statist. Plan. Inf.* **23**, 127–131.

Calvin, L.D. (1954). Doubly balanced incomplete block designs in which treatment effects are correlated, *Biometrics* **10**, 61–68.

Carmony, L.A. (1978). Tight $(2t + 1)$-designs, *Utilitas Math.* **14**, 39–47.

Cerenka, B. (1975). Affine resolvable incomplete block designs, *Zastowania Mat. Appl. Math.* **14**, 565–572.

Cerenka, B. (1976). Balanced block designs with unequal block sizes, *Biom. Z.* **18**, 499–504.

Chai, F.-S. and Majumdar, D. (1993). On the Yeh–Bradley conjecture on linear trend-free block designs, *Ann. Statist.* **21**, 2087–2097.

Chai, F.-S. and Majumdar, D. (2000). Optimal designs for nearest-neighbor analyses, *J. Statist. Plan. Inf.* **86**, 265–276.

Chai, F.-S. and Stufken, J. (1999). Trend-free block designs for higher order trends, *Utilitas Math.* **56**, 65–78.

Chakrabarti, M.C. (1963a). On the C-matrix in design of experiments, *J. Indian Statist. Assoc.* **1**, 8–23.

Chakrabarti, M.C. (1963b). On the use of incidence matrices of designs in sampling from finite populations, *J. Indian Statist. Assoc.* **1**, 78–85.

Chakravarti, I.M. and Ikeda, S. (1972). Construction of association schemes and designs from finite groups, *J. Comb. Th.* **13A**, 207–219.

Chang, L.-C. (1960). Association schemes for partially balanced designs with parameters $v = 28, n_1 = 12, n_2 = 15, p_{11}^2 = 4$, *Sci. Rec. New Ser.* **4**, 12–18.

Chang, Y. (2001). On the existence of BIB designs with large λ, *J. Statist. Plan. Inf.* **994**, 155–166.

Chaudhuri, A. and Mukerjee, R. (1988). *Randomized Response: Theory and Techniques*, Marcel Dekker, New York.

Cheng, C.-S. (1979). Optimal incomplete block designs with four varieties, *Sankhyā* **41B**, 1–14.

Cheng, C.-S. (1980). On the E-optimality of some block designs, *J. Roy. Statist. Soc.* **42B**, 199–204.

Cheng, C.-S. (1996). Optimal design: Exact theory, in *Handbook of Statistics*, 13 (eds. S. Ghosh and C.R. Rao), 977–1006. North-Holland, Amsterdom.

Cheng, C.-S. and Bailey, R.A. (1991). Optimality of some two-associate class partially balanced incomplete block designs, *Ann. Statist.* **19**, 1667–1671.

Cheng, C.-S. and Steinberg, D.M. (1991). Trend robust two-level factorial designs, *Biometrika* **76**, 245–251.

Cheng, C.-S. and Wu, C.-F.J. (1981). Nearly balanced incomplete block designs, *Biometrika* **68**, 493–500.

Clatworthy, W.H. (1955). Partially balanced incomplete block designs with two associate classes and two treatments per block, *J. Res. Nat. Bur. Stat.* **54**, 177–190.

Clatworthy, W.H. (1973). *Tables of Two-Associate Class Partially Balanced Designs, Applied Math. Ser. 63*, National Bureau of Standards, Washington, D.C.

Cochran, W.G. and Cox, G.M. (1957). *Experimental Designs*, 2nd edn. Wiley, New York.

Colbourn, C.J. and Dinitz, J.H. (eds.) (1996). *The CRC Handbook of Combinatorial Designs*, CRC Press, Boca Raton, Florida.

Collens, R.J. (1976). Constructing BIBD's with a computer, *Ars Comb.* **2**, 285–303.

Connife, D. and Stone, J. (1974). The efficiency factor of a class of incomplete block designs, *Biometrika* **61**, 633–636 (Correction: *Biometrika*, **62**, 686).

Connor, W.S. (1952). On the structure of balanced incomplete block designs, *Ann. Math. Statist.* **23**, 57–71.

Connor, W.S. and Clatworthy, W.H. (1954). Some theorems for partially balanced designs, *Ann. Math. Statist.* **25**, 100–112.

Constable, R.L. (1979). Some non-isomorphic solutions for the BIBD (12, 33, 11, 4, 3), *Utilitas Math.* **15**, 323–333.

Constantine, G.M. (1981). Some E-optimal block designs, *Ann. Statist.* **9**, 886–892.

Constantine, G.M. (1982). On the E-optimality of PBIB designs with a small number of blocks, *Ann. Statist.* **10**, 1027–1031.

Cornell, J.A. (1974). More on extended complete block designs. *Biometrics* **30**, 179–186.

Damaraju, L. and Raghavarao, D. (2002). Fractional plans to estimate main effects and two factor interactions inclusive of a specific factor, *Biom. J.* **44**, 780–785.

Das, M.N. (1954). On parametric relations in a balanced incomplete block design, *J. Indian Soc. Agric. Statist.* **6**, 147–152.

Das, M.N. and Narasimham, V.L. (1962). Construction of rotatable designs through BIB designs, *Ann. Math. Statist.* **33**, 1421–1439.

David, H.A. (1963). The structure of cyclic paired comparison designs, *J. Aust. Math. Soc.* **3**, 117–127.

David, H.A. (1967). Resolvable cyclic designs, *Sankhyā* **29A**, 191–198.

David, H.A. and Wolock, F.W. (1965). Cyclic designs, *Ann. Math. Statist.* **36**, 1526–1534.

David, O., Monod, H. and Amoussou, J. (2000). Optimal complete block designs to adjust for interplot competition with a covariance analysis. *Biometrics* **56**, 389–393.

Dehon, M. (1976a). Non-existence d' un 3-design de parameters $\lambda = 2$, $k = 5$ et $v = 11$, *Discr. Math.* **15**, 23–25.

Dehon, M. (1976b). Un theoreme d'extension de t-designs, *J. Comb. Th.* **21A**, 93–99.

Delsarte, P. (1976). Association schemes and t-designs in regular semi-lattices, *J. Comb. Th.* **20A**, 230–243.

Dènes, J. and Keedwell, A.D. (1974). *Latin Squares and Their Applications*, Academic Press, New York.

Denniston, R.H.F. (1976). Some new 5-designs, *Bull. London Math. Soc.* **8**, 263–267.

Desu, M.M. and Raghavarao, D. (2003). *Nonparametric Statistical Methods for Complete and Censored Data*, Chapman & Hall/CRC, Bota Racon, Florida.

Dey, A. (1970). On construction of balanced n-ary block designs, *Ann. Inst. Statist. Math.* **22**, 389–393.

Dey, A. (1977). Construction of regular group divisible designs. *Biometrika* **64**, 647–649.

Dey, A. (1986). *Theory of Block Designs*. Wiley Eastern, New Delhi.

Dey, A. and Midha, C.K. (1974). On a class of PBIB designs, *Sankhyā* **36B**, 320–322.

Dey, A. and Balasubramanian, K. (1991). Construction of some families of group divisible designs, *Utilitas Math.* **40**, 283–290.

Dey, A. and Saha, G.M. (1974). An inequality for tactical configurations, *Ann. Inst. Statist. Math.* **26**, 171–173.

Dey, A., Das, U.S. and Banerjee, A.K. (1986). Constructions of nested balanced incomplete block designs, *Calcutta Statist. Assoc. Bull.* **35**, 161–167.

Dey, A., Midha, C.K. and Trivedi, H.T. (1974). On construction of group divisible and rectangular designs, *J. Indian Soc. Agric. Statist.* **26**, 14–18.

Deza, M. (1975). There exist only a finite number of tight t-designs $s_\lambda(t, k, v)$ for every block size k, *Utilitas Math.* **8**, 347–348.

Ding, Y. and Raghavarao, D. (2005). Hadamard matrix methods in identifying differentially expressed genes from microarray experiments. Submitted to *J. Statist. Plan. Inf.*

Di Paola, J.W., Wallis, J.S. and Wallis, W.D. (1973). A list of (v, b, r, k, λ)-designs, $r < 30$, *Proc. 4th S.E. Conf. Combinatorics, Graph Th., and Computing*, 249–258.

Doob, M.S. (1972). On graph products and association schemes. *Utilitas Math.* **1**, 291–302.

Dorfman, R. (1943). The detection of defective members of a large population, *Ann. Math. Statist.* **14**, 436–440.

Downton, F. (1976). Non-parametric tests for block experiments. *Biometrika* **63**, 137–141.

Doyen, J. and Rosa, A. (1973). A bibliography and survey of Steiner systems, *Boll. U.M.I.* **7**, 392–419.

Doyen, J. and Rosa, A. (1977). An extended bibliography and Survey of Steiner Systems, *Proc. 7th Manitoba Conf. Numerical Math. Comput.*, 297–361.

Doyen, J. and Wilson, R.M. (1973). Embedding of Steiner triple systems, *Discr. Math.* **5**, 229–239.

Du, D.Z. and Hwang, F.K. (1993). *Combinatorial Group Testing and Its Applications*, World Scientific, Singapore.

Dudoit, S., Yang, Y.H., Callow, M.J. and Speed, T.P. (2002). Statistical methods for identifying differentially expressed genes in replicated cDNA microarray experiments. *Statist. Sin.* **12**, 111–139.

Dunnett, C.W. (1955). A multiple comparison procedure for comparing several treatments with control, *J. Amer. Statist. Assoc.* **50**, 1096–1121.

Eccleston, J.A. and Hedayat, A. (1974). On the theory of connected designs: Characterization and optimality, *Ann. Statist.* **2**, 1238–1255.

Edington, E.S. (1995). *Randomization Tests*, 3rd edn., Marcel Dekker, New York.

Ellenberg, J.H. (1977). The joint distribution of the standardized row sums of squares from a balanced two-way layout, *J. Amer. Statist. Assoc.* **72**, 407–411.

Federer, W.T. (1961). Augmented designs with one-way elimination of heterogeneity, *Biometrics* **17**, 447–473.

Federer, W.T. (1976). Sampling, blocking and model considerations for the completely randomized, randomized complete block, and incomplete block experimental designs, *Biom. Z.* **18**, 511–525.

Federer, W.T. (1993). *Statistical Design and Analysis for Intercropping Experiments. Two Crops*, Vol. I, Springer-Verlag, New York.

Federer, W.T. (1998). *Statistical Design and Analysis for Intercropping Experiments. Three or More Crops*, Vol. II, Springer-Verlag, New York.

Federer, W.T. and Raghavarao, D. (1975). On augmented designs, *Biometrics* **31**, 29–35.

Federer, W.T. and Raghavarao, D. (1987). Response models and minimal designs for mixtures of n of m items useful for intercropping and other investigations, *Biometrika* **74**, 571–577.

Federer, W.T., Joiner, J.R. and Raghavarao, D. (1974). Construction of orthogonal series 1 Balanced incomplete block designs — A new technique, *Utilitas Math.* **5**, 161–166.

Finizio, N.J. and Lewis, S.J. (1999). Some specializations of pitch tournament designs, *Utilitas Math.* **56**, 33–52.

Fisher, R.A. (1935–1951). *The Design and Analysis of Experiments*, Oliver & Boyd, Edinburgh.

Fisher, R.A. (1940). An examination of the different possible solutions of a problem in incomplete block designs, *Ann. Eugen.* **10**, 52–75.

Foody, W. and Hedayat, A. (1977). On theory and application of BIB designs with repeated blocks, *Ann. Statist.* **5**, 932–945.

Frank, D.C. and O'Shaughnessy, C.D. (1974). Polygonal designs, *Canad. J. Statist.* **2**, 45–59.

Freeman, G.H. (1976a). Incomplete block designs for three or four replicates of many treatments, *Biometrics* **32**, 519–528 (Correction: *Biometrics* **33**, 766).

Freeman, G.H. (1976b). A cyclic method of constructing regular group divisible incomplete block designs, *Biometrika* **63**, 555–558.

Friedman, M. (1937). The use of ranks to avoid the assumption of normality in the analysis of variance, *J. Amer. Statist. Assoc.* **32**, 675–701.

Fulton, J.D. (1976). PBIB designs with m associate classes induced by full rank matrices modulo p^m, *J. Comb. Th.* **20A**, 133–144.

Gaffke, N. (1982). D-optimal block designs with at most six varieties, *J. Statist. Plan. Inf.* **6**, 183–200.

Gardner, B. (1973). A new BIBD, *Utilitas Math.* **3**, 161–162.

Good, P. (2000). *Permutation Tests: A Practical Guide to Resampling Methods for Testing Hypotheses*, Springer-Verlag, New York.

Goos, P. (2002). *The Optimal Design of Blocked and Split-Plot Experiments.* Springer, New York.

Gower, J.C. and Preece, D.A. (1972). Generating successive incomplete blocks with each pair of elements in at least one block, *J. Comb. Th.* **12A**, 81–97.

Graver, J.E. (1972). A characterization of symmetric block designs, *J. Comb. Th.* **12A**, 304–308.

Griffiths, A.D. and Mavron, V.C. (1972). On the construction of affine designs, *J. London Math. Soc.* **5**, 105–113.

Grubbs, F.E. (1948). On estimating precision of measuring instruments and product variablility, *J. Amer. Statist. Soc.* **43**, 243–264.

Gupta, S.C. and Jones, B. (1983). Equireplicate balanced block designs with unequal block sizes, *Biometrika* **70**, 433–440.

Gupta, S.C. and Kageyama, S. (1992). Variance balanced designs with unequal block sizes and unequal replications, *Utilitas Math.* **42**, 15–24.

Gupta, V.K. (1993). Optimal nested block designs, *J. Indian Soc. Agric. Statist.* **45**, 187–194

Gupta, V.K., Parsad, R. and Das, A. (2003). Variance balanced block designs with unequal block sizes, *Utilitas Math.* **64**, 183–192.

Haemers, W. and Shrikhande, M.S. (1979). Some remarks on subdesigns of symmetric designs, *J. Statist. Plan. Inf.* **3**, 361–366.

Hall, M. Jr. (1967). *Combinatorial Theory*, Blaisdell Publishing Company, Waltham, MA.

Hall, M. Jr., Lane, R. and Wales, D. (1970). Designs derived from permutation groups, *J. Comb. Th.* **8**, 12–22.

Hall, W.B. and Jarrett, R.G. (1981). Non-resolvable incomplete block designs with few replications, *Biometrika* **68**, 617–627.

Hamada, N. (1973). On the p-rank of the incidence matrix of a balanced or partially balanced or partially balanced incomplete block designs and its applications in error correcting codes, *Hiroshima Math. J.* **3**, 153–226.

Hamada, N. (1974). On a geometrical method of construction of group divisible incomplete block designs, *Ann. Statist.* **2**, 1351–1356.

Hamada, N. and Kobayashi, Y. (1978). On the block structure of BIB designs with parameters $v = 22, b = 33, r = 12, k = 8$ and $\lambda = 4$, *J. Comb. Th.* **24A**, 75–83.

Hamada, N. and Tamari, F. (1975). Duals of balanced incomplete block designs derived from an affine geometry, *Ann. Statist.* **3**, 926–938.

Hamerle, A. (1978). A note on multiple comparison for randomized complete blocks with a binary response, *Biom. J.* **29**, 743–748.

Hanani, H. (1961). The existence and construction of balanced incomplete block designs, *Ann. Math. Statist.* **32**, 361–386.

Hanani, H. (1972). On balanced incomplete block designs with blocks having five elements, *J. Comb. Th.* **12A**, 184–201.

Hanani, H. (1974). On resolvable balanced incomplete block design, *J. Comb. Th.* **17A**, 275–289.

Hanani, H. (1975). Balanced incomplete block designs and related designs, *Discr. Math.* **11**, 255–369.

Hanani, H. (1979). A class of 3-design, *J. Comb. Th.* **26A**, 1–19.

Hanani, H., Ray-Chauduri, D.K. and Wilson, R.M. (1972). On irresolvable designs, *Discr. Math.* **3**, 343–357.

Harshberger, B. (1947). Rectangular lattices, *Virginia Agric. Expt. Statist. Res. Bull.* #347.

Harshberger, B. (1950). Triple rectangular lattices, *Biometrics* **5**, 1–13.

Hedayat, A. and Federer, W.T. (1974). Pairwise and variance balanced incomplete block designs, *Ann. Inst. Stat. Math.* **26**, 331–338.

Hedayat, A. and John, P.W.M. (1974). Resistant and susceptible BIB designs, *Ann. Statist.* **2**, 148–158.

Hedayat, A. and Kageyama, S. (1980). The family of t-designs — Part I, *J. Statist. Plan. Inf.* **4**, 173–212.

Hedayat, A.S. and Majumdar, D. (1985). Families of A-optimal block designs for comparing test treatments with a control, *Ann. Statist.* **13**, 757–767.

Hedayat, A. and Stufken, J. (1989). A relation between pairwise balanced and variance balanced block designs, *J. Amer. Statist. Assoc.* **84**, 753–755.

Herrendorfer, G. and Rasch, D. (1977). Complete block designs II. Analysis of partially balanced designs, *Biom. J.* **19**, 455–461.

Hinkelman, K. and Kempthorne, O. (1963). Two classes of group divisible partial diallel crosses, *Biometrika* **50**, 281–291.

Hirschfeld, J.W.P. (1979). *Projective Geometries Over Finite Fields*, Clarendon Press, Oxford.

Ho, Y.S. and Mendelsohn, N.S. (1973). Inequalities for t-designs with repeated blocks, *Aeq. Math.* **9**, 312–313.

Ho, Y.S. and Mendelsohn, N.S. (1974). Inequalities for t-designs with repeated blocks, *Aeq. Math.* **10**, 212–222.

Hochberg, Y. and Tamhane, A.C. (1987). *Multiple Comparisons Procedures*, Wiley, New York.

Hoffman, A.J. and Singleton, R.R. (1960). On Moore graphs with diameters 2 and 3, *I.B.M.J. Res. Dev.* **4**, 497–504.

Hormel, R.J. and Robinson, J. (1975). Nested partially balanced incomplete block designs, *Sankhyā* **37B**, 201–210.

Horvitz, D.G. and Thompson, D.J. (1952). A generalization of sampling without replacement from a finite universe, *J. Amer. Statist. Assoc.* **47**, 663–685.

Hubaut, X. (1974). Two new families of 4-designs, *Discr. Math.* **9**, 247–249.

Hughes, D.R. and Piper, F.C. (1976). On resolutions and Bose's theorem, *Geom. Ded.* **5**, 121–131.

Huynh, H. and Feldt, L.S. (1970). Conditions under which mean square ratios in repeated measurements designs have exact F-distributions, *J. Amer. Statist. Assoc.* **65**, 1582–1589.

Hwang, F.K. and Sos, V.T. (1981). Non-adaptive hypergeometric group testing, *Stud. Sci. Math. Hung.* **22**, 257–263.

Ionin, Y.J. and Shrikhande, M.S. (1998). Resolvable pairwise balanced designs, *J. Statist. Plan. Inf.* **72**, 393–405.

Ito, N. (1975). On tight 4-designs, *Osaka J. Math.* **12**, 493–522.

Jacroux, M. (1980). On the determination and construction of E-optimal block designs with unequal number of replicates, *Biometrika* **67**, 661–667.

Jacroux, M. (1983). Some minimum variance block designs for estimating treatment differences, *J. Roy. Statist. Soc.* **45B**, 70–76.

Jacroux, M. (1984). On the D-optimality of group divisible designs, *J. Statist. Plan. Inf.* **9**, 119–129.

Jacroux, M. (1985). Some sufficient conditions for the Type I optimality of block designs, *J. Statist. Plan Inf.* **11**, 385–398.

Jacroux, M., Majumdar, D. and Shah, K.R. (1995). Efficient block designs in the presence of trends, *Statist. Sin.* **5**, 605–615.

Jacroux, M., Majumdar, D. and Shah, K.R. (1997). On the determination and construction of optimal block designs in the presence of linear trends, *J. Amer. Statist. Assoc.* **92**, 375–382.

Jarrett, R.G. (1977). Bounds for the efficiency factor of block designs, *Biometrika* **64**, 67–72.

Jarrett, R.G. (1989). A review of bounds for the efficiency factor of block designs, *Aust. J. Statist.* **31**, 118–129.

Jarrett, R.G. and Hall, W.B. (1978). Generalized cyclic incomplete block designs, *Biometrika* **65**, 397–401.

Jimbo, M. and Kuriki, S. (1983). Constructions of nested designs, *Ars Comb.* **16**, 275–285.

John, J.A. (1966). Cyclic incomplete block designs, *J. Roy. Statist. Soc.* **28B**, 345–360.

John, J.A. (1987). *Cyclic Designs*, Chapman & Hall, London.

John, J.A. and Mitchell, T.J. (1977). Optimal incomplete block designs, *J. Roy. Statist. Soc.* **39B**, 39–43.

John, J.A. and Turner, G. (1977). Some new group divisible designs, *J. Statist. Plan. Inf.* **1**, 103–107.

John, J.A., Wolock, F.W. and David, H.A. (1972). *Cyclic Designs. Applied Math. Ser.* 62, National Bureau of Standards, Washington, D.C.

John, P.W.M. (1962). Testing two treatments when there are three experimental units in each block, *Appl. Statist.* **11**, 164–169.

John, P.W.M. (1966). An extension of the triangular association scheme to three associate classes, *J. Roy. Statist. Soc.* **28B**, 361–365.

John, P.W.M. (1970). The folded cubic association scheme, *Ann. Math. Statist.* **41**, 1983–1991.

John, P.W.M. (1973). A balanced design for 18 varieties, *Technometrics* **15**, 641–642.

John, P.W.M. (1975). Isomorphic $L_2(7)$ designs, *Ann. Statist.* **3**, 728–729.

John, P.W.M. (1976a). Robustness of balanced incomplete block deigns, *Ann. Statist.* **4**, 960–962.

John, P.W.M. (1976b). Inequalities for semi-regular group divisible designs, *Ann. Statist.* **4**, 956–959.

John, P.W.M. (1977a). Series of semi-regular group divisible designs, *Comm. Statist.* **6A**, 1385–1392.

John, P.W.M. (1977b). Non-isomorphic differences sets for $v = 31$, *Sankhyā* **39B**, 292–293.

John, P.W.M. (1978a). A new balanced design for eighteen varieties, *Technometrics* **20**, 155–158.

John, P.W.M. (1978b). A note on concordance matrices for partially balanced designs, *Utilitas Math.* **14**, 3–7.

John, P.W.M. (1978c). A note on a finite population plan, *Ann. Statist.* **6**, 697–699.

John, P.W.M. (1980). *Incomplete Block Designs*, Lecture Notes in Statistics 1, Mercel Dekker, New York.

Jones, B., Sinha, K. and Kageyama, S. (1987). Further equireplicate variance balanced designs with unequal block sizes, *Utilitas Math.* **32**, 5–10.

Jones, R.M. (1959). On a property of incomplete blocks, *J. Roy. Statist. Soc.* **21B**, 172–179.

Kageyama, S. (1971). An improved inequality for balanced incomplete block designs, *Ann. Math. Statist.* **42**, 1448–1449.

Kageyama, S. (1972a). A survey of resolvable solutions of balanced incomplete block designs, *Int. Statist. Rev.* **40**, 269–273.

Kageyama, S. (1972b). Note on Takeuchi's table of difference sets generating balanced incomplete block designs, *Int. Statist. Rev.* **40**, 275–276.

Kageyama, S. (1972c). On the reduction of associate classes for certain PBIB designs, *Ann. Math. Statist.* **43**, 1528–1540.

Kageyama, S. (1973a). On the inequality for BIBDS with special parameters, *Ann. Statist.* **1**, 204–207.

Kageyama, S. (1973b). On μ-resolvable and affine μ-resolvable balanced incomplete block designs, *Ann. Statist.* **1**, 195–203.

Kageyama, S. (1973c). A series of 3-designs, *J. Jpn. Statist. Soc.* **3**, 67–68.

Kageyama, S. (1974a). On the incomplete block designs derivable from the association schemes, *J. Comb. Th.* **17A**, 269–272.

Kageyama, S. (1974b). Note on the reduction of associate classes for PBIB designs, *Ann. Inst. Statist. Math.* **26**, 163–170.

Kageyama, S. (1974c). Reduction of associate classes for block designs and related combinatorial arrangements, *Hiroshima Math. J.* **4**, 527–618.

Kageyama, S. (1974d). On the reduction of associate classes for the PBIB design of a general type, *Ann. Statist.* **2**, 1346–1350.

Kageyama, S. (1975). Note on an inequality for tactical configurations, *Ann. Inst. Statist. Math.* **27**, 529–530.

Kageyama, S. (1976). Constructions of balanced block designs, *Utilitas Math.* **9**, 209–229.

Kageyama, S. (1976). Resolvability of block designs, *Ann. Statist.* **4**, 655–661.

Kageyama, S. (1977a). Classification of certain t-designs, *J. Comb. Th.* **22A**, 163–168.

Kageyama, S. (1977b). Comparison of the bounds for the block intersection number of incomplete block designs, *Calcutta Statist. Assoc. Bull.* **26**, 53–60.

Kageyama, S. (1978a). Bounds on the number of disjoint blocks of incomplete block designs, *Utilitas Math.* **14**, 177–180.

Kageyama, S. (1978b). Improved inequalities for balanced incomplete block designs, *Discr. Math.* **21**, 177–188.

Kageyama, S. (1979). Mathematical expression of an inequality for a block design, *Ann. Inst. Statist. Math.* **31**, 293–298.

Kageyama, S. (1980). On properties of efficiency balanced designs, *Comm. Statist.* **9A**, 597–616.

Kageyama, S. (1989). Existence of variance balanced binary design with fewer experimental units, *Statist. Prob. Lett.* **7**, 27–28.

Kageyama, S. and Hedayat, A.S. (1983). The family of t-designs. Part II, *J. Statist. Plan. Inf.* **7**, 257–287.

Kageyama, S. and Ishii, G. (1975). A note on BIB designs, *Aust. J. Statist.* **17**, 49–50.

Kageyama, S. and Miao, Y. (1997). Nested designs with block size five and subblock size two, *J. Statist. Plan. Inf.* **64**, 125–139.

Kageyama, S. and Nishii, R. (1980). Non-existence of tight $(2s + 1)$ designs, *Utilitas Math.* **17**, 225–229.

Kageyama, S. and Tsuji, T. (1977a). Characterization of certain incomplete block designs, *J. Statist. Plan. Inf.* **1**, 151–161.

Kageyama, S. and Tsuji, T. (1977b). General upperbound for the number of blocks having a given number of treatments common with given block, *J. Statist. Plan. Inf.* **1**, 217–234.

Kageyama, S. and Tsuji, T. (1979). Inequality for equi-replicated n-ary block designs with unequal block sizes, *J. Statist. Plan. Inf.* **3**, 101–107.

Kageyama, S., Philip, J. and Banerjee, S. (1995). Some constructions of nested BIB and two-associate PBIB designs under restricted dualization, *Bull. Fac. Sch. Educ. Hiroshima Uviv.* Part II, **17**, 33–39.

Kageyama, S., Saha, G.M. and Das, A (1978). Reduction of the number of classes of hypercubic association schemes, *Ann. Inst. Statist. Math.* **30**, 115–123.

Kantor, W.M. (1969). 2-transitive symmetric designs, *Trans. Amer. Math. Soc.* **146**, 1–28.

Kempthorne, O. (1953). A class of experimental designs using blocks of two plots, *Ann. Math. Statist.* **24**, 76–84.

Kempthorne, O. (1956). The efficiency factor of an incomplete block design, *Ann. Math. Statist.* **27**, 846–849.

Khatri, C.G. (1982). A note on variance balanced designs, *J. Statist. Plan. Inf.* **6**, 173–177.

Kiefer, J.C. (1975a). Balanced block designs and generalized Youden designs. I. Construction (patchwork), *Ann. Statist.* **3**, 109–118.

Kiefer, J.C. (1975b). Construction and optimality of generalized Youden designs, in *A Survey of Statistical Design and Linear Model* (ed. J.N. Srivastava) 333–353. North-Holland, Amsterdam.

Kim, K. and Stufken, J. (1995). On optimal block designs for comparing a standard treatment to test treatments, *Utilitas Math.* **47**, 211–224.

Kingsley, R.A. and Stantion, R.G. (1972). A survey of certain balanced incomplete designs, *Proc. 3rd S.E. Conf. Combinatorics, Graph, Th. and Computing*, 305–310.

Kirkman, T.P. (1847). On a problem of combinations, *Cambridge and Dublin Math J.* **2**, 191–204.

Kirkman, T.P. (1850). Query VI on "Fifteen young ladies…". *Lady's and Gentleman's Diary*, No. 147, 48.

Kishen, K. (1940–1941). Symmetrical unequal block arrangements, *Sankhyā* **5**, 329–344.

Koukouvinos, C., Koumias, S. and Seberry, J. (1989). Further Hadamard matrices with maximal excess and new SBIBD $(4k^2, 2k^2 + k, k^2)$, *Utilitas Math.* **36**, 135–150.

Kramer, E.S. (1975). Some t-designs for $t > 4$ and $v = 17, 18$, *Proc. 6th S.E. Conf. Combinatorics, Graph Th., and Computing*, 443–460.

Kramer, E.S. and Mesner, D.M. (1975). Admissible parameters for Steiner systems $S(t, k, v)$ with a table for all $(v - t) \leq 498$, *Utilitas Math.* **7**, 211–222.

Kreher, D. (1996). t-designs, $t \geq 3$, in *Handbook of Combinatorial Designs* (eds. J. Dinitz and C. Colburn) 47–66. CRC Press, Bota Racon, Florida.

Kulshreshtha, A.C. (1972). A new incomplete block design for slope-ratio assays, *Biometrics* **28**, 585–589.

Kulshreshta, A.C., Dey, A. and Saha, G.M. (1972). Balanced designs with unequal replications and unequal block sizes, *Ann. Math. Statist.* **43**, 1342–1345.

Kusumoto, K. (1965). Hyper cubic designs, *Wakayama Med. Rep.* **9**, 123–132.

Kusumoto, K. (1967). Association schemes of new types and necessary conditions for existence of regular and symmetrical PBIB designs with those association scheme, *Ann. Inst. Statist. Math.* **19**, 73–100.

Kusumoto, K. (1975). Necessary and sufficient condition for the reducibility of association schemes, *J. Jpn. Statist. Soc.* **5**, 57–64.

Lander, E.S. (1983). *Symmetric designs: An algebraic approach*, London Mathematical Society Lecture Notes Series 74, Cambridge University Press, Cambridge.

Lawless, J.F. (1970). Block intersections in quasi-residual designs, *Aeq. Math.* **5**, 40–46.

Li, P.C. and Van Rees, G.H.J. (2000). New construction of lotto designs, *Utlitas Math.* **58**, 45–64.

Lindner, C.C. (1976). Embedding block designs into resolvable block designs, *Ars Comb.* **1**, 215–219.

Lindner, C.C. (1978). Embedding block designs into resolvable block designs II, *Ars Comb.* **6**, 49–55.

Macula, A.J. and Rykov, V.V. (2001). Probabilistic nonadaptive group testing in the presence of errors and inhibitors, in *Recent Advances in Experimental Designs and Related Topics* (eds. S. Altan and J. Singh) 73–86, Nova Science Publishers, New York.

Magda, C.G. (1979). On E-optimal designs and Schur optimality, Ph.D. Thesis, University of Illinois at Chicago.

Majindar, K.N. (1978). Intersection of blocks in a quasi-residual balanced incomplete block design, *Utilitas Math.* **13**, 189–191.

Majumdar D. (1996). On admissibility and optimality of treatment-control designs, *Ann. Statist.*, **24**, 2097–2107.

Majumdar, D. and Notz, W.I. (1983). Optimal incomplete block designs for comparing treatments with a control, *Ann. Statist.* **11**, 258–266.

Maloney, C.J. (1973). Grubb's estimator to date, in *Proceedings of the 18th Conference on Design of Experiments in Army Research, Development and Testing*, Part 2, 523–547.

Mann, H.B. (1965). *Addition Theorems: The Addition Theorems of Group Theory and Number Theory*, John Wiley, New York.

Mann, H.B. (1969). A note on balanced incomplete block designs, *Ann. Math. Statist.* **40**, 679–680.

Martin, R.J. and Eccleston, J.A. (1991). Optimal incomplete block designs for general dependence structures, *J. Statist. Plan. Inf.* **28**, 67–81.

McCarthy, P.J. (1969). Pseudo-replication: Half samples, *Int. Statist. Rev.* **33**, 239–264.

Mehta, S.K., Agarwal, S.K. and Nigam, A.K. (1975). On partially balanced incomplete block designs through partially balanced ternary designs, *Sankhyā* **37B**, 211–219.

Mejza, S. and Cerenka, B. (1978). On some properties of the dual of a totally balanced block design, *J. Statist. Plan. Inf.* **2**, 93–94.

Mendelsohn, N.S. (1970). A theorem on Steiner systems, *Canad. J. Math.* **22**, 1010–1015.

Mendelsohn, N.S. (1978). Perfect cyclic designs, *Disc. Math.* **20**, 63–68.

Mesner, D.M. (1967). A new family of partially balanced incomplete block designs with some Latin square design properties, *Ann. Math. Statist.* **38**, 571–581.

Mills, W.H. (1975). Two new block designs, *Utilitas Math.* **7**, 73–75.

Mills, W.H. (1977). The construction of balanced incomplete block designs with $\lambda = 1$, *Proc. 7th Manitoba Conf. Numerical Math. Comput.* 131–148.

Mills, W.H. (1978). A new 5-design, *Ars Comb.* **6**, 193–195.

Mitchell, T.J. and John, J.A. (1976). Optimal incomplete block designs, Oak Ridge National Laboratory/CSD-8.

Mohan, R.N., Kageyama, S. and Nair, M.M. (2004). On a characterization of symmetric balanced incomplete block designs, *Disc. Math. Prob. Statist.* **24**, 41–58.

Mood, A.M. (1946). On Hotelling's weighing problem, *Ann. Math. Statist.* **17**, 432–446.

Morales, L.B. (2002). Constructing some PBIBD (2)s by Tabu search algorithm, *JCMCC* **43**, 65–82.

Moran, M.A. (1973). The analysis of incomplete block designs as used in the group screening of drugs, *Biometrics* **29**, 131–142.

Morgan, E.J. (1977). Construction of balanced n-ary designs, *Utilitas Math.* **11**, 3–31.

Morgan, J.P. and Uddin, N. (1995). Optimal, non-binary variance balanced designs, *Statist. Sin.* **5**, 535–546.

Morgan, J.P., Preece, D.A. and Rees, D.H. (2001). Nested balanced incomplete block designs, *Disc. Math.* **231**, 351–389.

Most, B.M. (1976). Resistance of balanced incomplete block designs, *Ann. Statist.* **3**, 1149–1162.

Mukerjee, R. and Kageyama, S. (1985). On resolvable and affine resolvable variance balanced designs, *Biometrika* **72**, 165–172.

Mukhopadhyay, A.C. (1972). Construction of BIBD's from OA's and combinatorial arrangements analogous to OA's, *Calcutta Statist. Assoc. Bull.* **21**, 45–50.

Mukhopadhyay, A.C. (1974). On constructing balanced incomplete block designs from association matrices, *Sankhyā* **36B**, 135–144.

Mukhopadhyay, A.C. (1978). Incomplete block designs through orthogonal and incomplete orthogonal arrays, *J. Statist. Plan. Inf.* **2**, 189–195.

Mullin, R.C. (1974). Resolvable designs and geometroids, *Utilitas Math.* **5**, 137–149.

Mullin, R.C. and Collens, R.J. (1974). New block designs and BIBD's, *Proc. 5th S.E. Conf. Combinatories, Graph Th., and Computing* 345–366.

Murty, J.S. and Das, M.N. (1968). Balanced n-ary block designs and their uses, *J. Indian Statist. Assoc.* **5**, 73–82.

Myers, R.H. and Montgomery, D.C. (1995). *Response Surface Methodology*, Wiley, New York.

Nair, C.R. (1964). A new class of designs, *J. Amer. Statist. Assoc.* **59**, 817–833.

Nair, C.R. (1966). On partially linked block designs, *Ann. Math. Statist.* **37**, 1401–1406.

Nair, C.R. (1980). Construction of PBIB designs from families of difference sets, *J. Indian Soc. Agric. Statist.* **32**, 33–40.

Nair, K.R. (1951). Rectangular lattices and partially balanced incomplete block designs, *Biometrics* **7**, 145–154.

Nash-Williams, C. St. J.A. (1972). Simple constructions for balanced incomplete block designs with block size 3, *J. Comb. Th.* **13A**, 1–6.

Nelder, J.A. (1965). The analysis of randomized experiments with orthogonal block structures (Parts I and II), *Proc. Roy. Soc.* **283A**, 147–178.

Nguyen, N.-K. (1990). Computer aided construction of small (M, S)-optimal incomplete block designs, *Aust. J. Statist.* **32**, 399–410.

Nguyen, N.-K. (1993). An algorithm for constructing optimal resolvable incomplete block designs, *Comm. Statist.* **22B**, 911–923.

Nguyen, N.-K. (1994). Construction of optimal block designs by computer, *Technometrics* **36**, 300–307.

Nigam, A.K. (1974a). A note on construction of triangular PBIB design with parameters $v = 21, b = 35, r = 10, k = 6, \lambda_1 = 2, \lambda_2 = 3$, *J. Indian Soc. Agric. Statist.* **26**, 46–47.

Nigam, A.K. (1974b). Construction of balanced n-ary block designs and partially balanced arrays, *J. Indian Soc. Agric. Statist.* **26**, 48–56.

Nigam, A.K. (1976). Nearly balanced incomplete block designs. *Sankhyā* **38B**, 195–198.

Nigam, A.K., Mehta, S.K. and Agarwal, S.K. (1977). Balanced and nearly balanced n-ary designs with varying block sizes and replications, *J. Indian Soc. Agric. Statist.* **29**, 92–96.

Nigam, A.K., Puri, P.D. and Gupta, V.K. (1988). *Characterizations and Analysis of Block Designs*, Wiley Eastern, New Delhi.

Noda, R. (1976). Some inequalities for t-designs, *Osaka J. Math.* **13**, 1361–1366.

Noda, R. (1978). On some $t - (2k, k, \lambda)$ designs, *J. Comb. Th.* **24A**, 89–95.

Notz, W.I. and Tamhane, A.C. (1983). Balanced treatment incomplete block (BTIB) designs for comparing treatments with a control: Minimal complete sets of generator designs for $k = 3, p = 3(1)10$, *Comm. Statist.* **12B**, 1391–1412.

Ogasawara, M. (1965). A necessary condition for the existence of regular and symmetrical PBIB designs of T_m type, *Inst. Statist. Memeo Ser.* 418.

Ogawa, J. and Ikeda, S. (1973). On the randomization of block designs. in *A Survey of Combinatorial Theory* (eds. J.N. Srivastava *et al.*), 335–347.

Paik, U.B. and Federer, W.T. (1973a). Partially balanced designs and properties A and B, *Comm. Statist.* **1**, 331–350.

Paik, U.B. and Federer, W.T. (1973b). On PA type incomplete block designs and PAB type rectangular designs, *J. Roy. Statist. Soc.* **35B**, 245–251.

Paik, U.B. and Federer, W.T. (1977). Analysis of binary number association scheme partially balanced designs, *Comm. Statist.* **6A**, 895–931.

Parker, E.T. (1963). Remarks on balanced incomplete block designs, *Proc. Ann. Math. Soc.* **14**, 729–730.

Parker, E.T. (1975). On sets of pairwise disjoint blocks in Steiner triple systems, *J. Comb. Th.* **19A**, 113–114.

Paterson, L.J. (1983). Circuits and efficiency factors in incomplete block designs, *Biometrika* **70**, 215–226.

Patterson, H.D. and Thompson, R. (1971). Recovery of inter-block information when block sizes are unequal, *Biometrika* **58**, 545–554.

Patterson, H.D. and Williams, E.R. (1976a). A new class of resolvable incomplete block designs, *Biometrika* **63**, 83–92.

Patterson, H.D. and Williams, E.R. (1976b). Some theoretical results on general block designs, *Proc. 5th Br. Combinatorial Conf. Congressus Numerentium*, **XV**, 489–496.

Patterson, H.D., Williams, E.R. and Hunter, E.A. (1978). Block designs for variety trials, *J. Agric. Sci.* **90**, 395–400.

Patwardhan, G.A. (1972). On identical blocks in a balanced incomplete block design, *Calcutta Statist. Assoc. Bull.* **21**, 197–200.

Pearce, S.C. (1960). Supplemented balance, *Biometrika* **47**, 263–271.

Pearce, S.C. (1963). The use and classification of non-orthogonal designs, *J. Roy. Statist. Soc.* **126A**, 353–377.

Pearce, S.C. (1964). Experimenting with blocks of natural size, *Biometrics* **20**, 699–706.

Pearce, S.C. (1968). The mean efficiency of equi-replicated designs, *Biometrika* **55**, 251–253.

Pearce, S.C. (1970). The efficiency of block designs in general, *Biometrika* **57**, 339–346.

Pearce, S.C. (1971). Precision in block experiments, *Biometrika* **58**, 161–167.

Pesuk, J. (1974). The efficiency of controls in balanced incomplete block designs, *Biom. Z.* **16**, 21–26.

Peterson, C. (1977). On tight t-designs, *Osaka J. Math.* **14**, 417–435.

Preece, D.A. (1967a). Nested balanced incomplete block designs, *Biometrika* **54**, 479–486.

Preece, D.A. (1967b). Incomplete block designs with $v = 2k$, *Sankhyā* **29A**, 305–316.

Preece, D.A. (1977). Orthogonality and designs: A terminology muddle, *Utilitas Math.* **12**, 201–223.

Preece, D.A. (1982). Balance and designs: Another terminological tangle, *Utilitas Math.* **21**, 85–186.

Prescott, P. and Mansson, R. (2001). Robustness of balanced incomplete block designs to randomly missing observations, *J. Statist. Plan. Inf.* **92**, 283–296.

Puri, P.D. and Nigam, A.K. (1975a). On patterns of efficiency balanced designs, *J. Roy Statist. Soc.* **37B**, 457–458.

Puri, P.D. and Nigam, A.K. (1975b). A notion of efficiency balanced designs, *Sankhyā* **37B**, 457–460.

Puri, P.D. and Nigam, A.K. (1977a). Partially efficiency balanced designs, *Comm. Statist.* **6A**, 753–771.

Puri, P.D. and Nigam, A.K. (1977b). Balanced block designs, *Comm. Statist.* **6A**, 1171–1179.

Puri, P.D., Nigam, A.K. and Narain, P. (1977c). Supplemented block designs, *Sankhyā* **39B**, 189–195.

Raghavachari, M. (1977). Some properties of the width of the incidence matrices of balanced incomplete block designs and Steiner triple systems, *Sankhyā* **39B**, 181–188.

Raghavarao, D. (1960a). On the block structure of certain PBIB designs with triangular and L_2 association schemes, *Ann. Math. Statist.* **31**, 787–791.

Raghavarao, D. (1960b). A generalization of group divisible designs, *Ann. Math. Statist.* **31**, 756–771.

Raghavarao, D. (1962a). Symmetrical unequal block arrangements with two unequal block sizes, *Ann. Math. Statist.* **33**, 620–633.

Raghavarao, D. (1962b). On balanced unequal block designs, *Biometrika* **49**, 561–562.

Raghavarao, D. (1963). On the use of latent vectors in the analysis of group divisible and L_2 designs, *J. Indian Soc. Agric. Statist.* **14**, 138–144.

Raghavarao, D. (1971). *Constructions and Combinatorial Problems in Design of Experiments*, Wiley, New York (Dover Edition, 1988).

Raghavarao, D. (1973). Method of differences in the construction of $L_2(s)$ designs, *Ann. Statist.* **1**, 591–592.

Raghavarao, D. (1974). Boolean sums of sets of certain designs, *Disc. Math.* **10**, 401–403.

Raghavarao, D. and Aggarwal, K.R. (1973). Extended generalized right angular designs and their applications as confounded asymmetrical factorial experiments, *Calcutta Statist. Assoc. Bull.* **22**, 115–120.

Raghavarao, D. and Aggarwal, K.R. (1974). Some new series of PBIB designs and their applications, *Ann. Inst. Statist. Math.* **26**, 153–161.

Raghavarao, D. and Chandrasekhararao, K. (1964). Cubic designs, *Ann. Math. Statist.* **35**, 389–397.

Raghavarao, D. and Federer, W.T. (1979). Block total response as an alternative to randomized response method in survey, *J. Roy. Statist. Soc.* **41B**, 40–45.

Raghavarao, D. and Federer, W.T. (2003). Sufficient conditions for balanced incomplete block designs to be minimal fractional combinatorial treatment designs, *Biometrika* **90**, 465–470.

Raghavarao, D. and Rao, G.N. (2001). Design and analysis when the intercrops are in differenct classes, *J. Indian Soc. Agric Statist.* **54**, 236–243.

Raghavarao, D. and Singh, R. (1975). Applications of PBIB designs in cluster sampling, *Proc. Indian Nat. Sci. Acad.* **41**, 281–288.

Raghavarao, D. and Wiley, J.B. (1986). Testing competing effects among soft drink brands, in *Statistical Design: Theory and Practice* (eds. C.E. McCullock *et al.*), 161–167. Cornell University Press, Ithaca.

Raghavarao, D. and Wiley, J.B. (1998). Estimating main effects with Pareto optimal subsets, *Aust. New Zealand J. Statist.* **40**, 425–432.

Raghavarao, D. and Zhang, D. (2002). 2^n behavioral experiments using Pareto optimal choice sets, *Statist. Sin.* **12**, 1085–1092.

Raghavarao, D. and Zhou, B. (1997). A method of constructing 3-designs, *Utilitas Math.* **52**, 91–96.

Raghavarao, D. and Zhou, B. (1998). Universal optimality of UE 3-designs for a competing effects model, *Comm. Statist.* **27A**, 153–164.

Raghavarao, D., Federer, W.T. and Schwager, S.J. (1986). Characteristics for distinguishing balanced incomplete block designs with repeated blocks, *J. Statist. Plan. Inf.* **13**, 151–163.

Raghavarao, D., Shrikhande, S.S. and Shrikhande, M.S. (2002). Incidence matrices and inequalities for combinatorial designs, *J. Comb. Designs* **10**, 17–26.

Ralston, T. (1978). Construction of PBIBD's, *Utilitas Math.* **13**, 143–155.

Rao, C.R. (1947). General methods of analysis for incomplete block designs, *J. Amer. Statist. Assoc.* **42**, 541–561.

Rao, C.R. (1973). *Linear Statistical Inference and Its Applications*, Wiley, New York.

Rao, J.N.K. and Nigam, A.K. (1990). Optimal controlled sampling designs, *Biometrika* **77**, 807–814.

Rao, J.N.K. and Vijayan, K. (2001). Applications of experimental designs in survey sampling, in *Recent Advances in Experimental Designs and Related Topics* (eds. S. Altan and J. Singh), 65–72, Nova Science Publishers, Inc., New York.

Rao, V.R. (1958). A note on balanced designs, *Ann. Math. Statist.* **29**, 290–294.

Rasch, D. (1978). Selection problems in balanced block designs, *Biom. J.* **20**, 275–278.

Rasch, D. and Herrendörfer, G. (1976). Complete block designs. I. Models, definitions and analysis of balanced incomplete block designs, *Biom. J.* **18**, 609–617.

Ray-Chaudhuri, D.K. and Wilson, R.M. (1971). Solution of Kirkman's school girl's problem, *Proc. Symp. Math. XIX*, American Mathematical Society, Providence, RI **19**, 187–204.

Ray-Chaudhuri, D.K. and Wilson, R.M. (1973). The existence of resolvable blocks designs, in *A Survey of Combinatorial Theory* (eds. J.N. Srivastava *et al.*), 361–375.

Ray-Chauduri, D.K. and Wilson, R.M. (1975). On t-designs, *Osaka J. Math.* **12**, 737–744.

Robinson, J. (1975). On the test for additivity in a randomized block design, *J. Amer. Statist. Assoc.* **70**, 184–185.

Rosa, A. (1975). A theorem on the maximum number of disjoint Steiner triple systems, *J. Comb. Th.* **18A**, 305–312.

Roy, J. (1958). On the efficiency factor of block designs, *Sankhyā* **19**, 181–188.

Roy, J. and Laha, R.G. (1956). Classification and analysis of linked block designs, *Sankhyā* **17**, 115–132.

Roy, J. and Laha, R.G. (1957). On partially balanced linked block designs, *Ann. Math. Statist.* **28**, 488–493.

Roy, S.N. and Srivastava, J.N. (1968). Inference on treatment effects in incomplete block designs, *Rev. Int. Statist. Inst.* **36**, 1–6.

Russell, K.G. and Eccleston, J.A. (1987a). The construction of optimal balanced incomplete block designs when adjacent observations are correlated, *Aust. J. Statist.* **29**, 77–89.

Russel, K.G. and Eccleston, J. (1987b). The construction of optimal incomplete block designs when observations within a block are correlated. *Austr. J. Statist.* **29**, 293–302.

Russell, T.S. and Bradley, R.A. (1958). One way variance in a two-way classification. *Biometrika* **45**, 111–129.

Saha, G.M. (1973). On construction of T_m-type PBIB designs, *Ann. Inst. Statist. Math.* **25**, 605–616.

Saha, G.M. (1975a). Some results in tactical configurations and related topics, *Utilitas Math.* **7**, 223–240.

Saha, G.M. (1975b). On construction of balanced ternary designs, *Sankhyā* **37**, 220–227.

Saha, G.M. (1976). On Calinski's paterns in block designs, *Sankhyā* **38B**, 383–392.

Saha, G.M. (1978). A note on combining two partially balanced incomplete block designs, *Utilitas Math.* **13**, 27–32.

Saha, G.M. and Das, A. (1978). An infinite class of PBIB designs, *Canad. J. Statist.* **6**, 25–32.

Saha, G.M. and Das, M.N. (1971). Construction of partially balanced incomplete block designs through 2^n factorials and some new designs of two associate classes. *J. Comb. Th.* **11A**, 282–295.

Saha, G.M. and Dey, A. (1973). On construction and uses of balanced n-ary designs, *Ann. Inst. Statist. Math.* **25**, 439–445.

Saha, G.M. and Shanker, G. (1976). On a generalized group divisible family of association schemes and PBIB designs based on the schemes, *Sankhyā* **38B**, 393–404.

Saha, G.M., Dey, A. and Kulshrestha, A.C. (1974). Circular designs — Further results, *J. Indian Soc. Agric. Statist.* **26**, 87–92.

Saha, G.M., Kulshreshtha, A.C. and Dey, A. (1973). On a new type of m-class cyclic association scheme and designs based on the scheme, *Ann. Statist.* **1**, 985–990.

Schaefer, M. (1979). On the existence of binary, variance balanced block designs, *Utilitas Math.* **16**, 87–95.

Scheffè, H. (1959). *The Analysis of Variance*. John Wiley, New York.

Schellenberg, P.J., Van Rees, G.H.J. and Vanstone, S.A. (1977). The existence of balanced tournament designs, *Ars Comb.* **3**, 303–318.

Schreckengost, J.F. and Cornell, J.A. (1976). Extended complete block designs with correlated observations, *Biometrics* **32**, 505–518.

Schreiber, S. (1974). Some balanced incomplete block designs, *Israel J. Math.* **18**, 31–37.

Schultz, D.J., Parnes, M. and Srinivasan, R. (1993). Further applications of d-complete designs in group testing, *J. Comb. Inf. Syst. Sci.* **18**, 31–41.

Seberry, J. (1978). A class of group divisible designs, *Ars Comb.* **6**, 151–152.

Seberry, J. (1991). Existence of SBIBD $(4k^2, 2k^2+k, k^2 + k)$ and Hadamand matrices with maximal excess, *Aust. J. Comb.* **4**, 87–91.

Seiden, E. (1963). A supplement to Parker's "Remarks on balanced incomplete block designs", *Proc. Amer. Math. Soc.* **14**, 731–732.

Seiden, E. (1977). On determination of the structure of a BIB design (7, 21, 9, 3, 3) by the number of distinct blocks. Technical Report RM 376, Department of Statistics, Michigan State Universit East Lansing.

Sethuraman, V.S., Raghavarao, D. and Sinha, B.K. (2005). Optimal s^n factorial designs when observations within blocks are correlated, *Comp. Statist. Data Anal.*, in press.

Shafiq, M. and Federer, W.T. (1979). Generalized N-ary balanced block designs, *Biometrika* **66**, 115–123.

Shah, B.V. (1958). On balancing in factorial experiments, *Ann. Math. Statist.* **29**, 766–779.

Shah, B.V. (1959a). A generalization of partially balanced incomplete block designs, *Ann. Math. Statist.* **30**, 1041–1050.

Shah, B.V. (1959b). A note on orthogonality in experimental designs, *Calcutta Statist. Assoc. Bull.* **8**, 73–80.

Shah, K.R. and Sinha, B.K. (1989). *Theory of Optimal Designs*. Springer Lecture Notes Series 54, New York.

Shah, K.R. and Sinha, B.K. (2001). Discrete optimal designs: Criteria and characterization, in *Recent Advances in Experimental Designs and Related Topics* (eds. S. Altan and J. Singh), 133–152, Nova, New York.

Shah, K.R., Raghavarao, D. and Khatri, C.G. (1976). Optimality of two and three factor designs, *Ann. Statist.* **4**, 419–422.

Shah, S.M. (1975a). On some block structure properties of a BIBD, *Calcutta Statist. Assoc. Bull.* **24**, 127–130.

Shah, S.M. (1975b). On the existence of affineg μ-resolvable BIBDS. *Applied Statistics* (ed. R.P. Gupta), North-Holland, 289–293.

Shah, S.M. (1976). Bounds for the number of identical blocks in incomplete block designs, *Sankhyā* **38B**, 80–86.

Shah, S.M. (1977). An upper bound for the number of treatments common to any set of blocks of certain PBIB designs, *Sankhyā* **39B**, 76–83.

Shah, S.M. (1978). On α-resolvability of incomplete block designs, *Ann. Statist.* **6**, 468–471.

Shao, J. (1993). Linear model selection by cross-validation, *J. Amer. Statist. Assoc.* **88**, 486–494.

Sharma, H.C. (1978). On the construction of some L_2 designs, *Calcutta Statist. Assoc. Bull.* **27**, 81–96.

Sharma, S.D. and Agarwal, B.L. (1976). Some aspects of balanced n-ary designs, *Sankhyā* **38B**, 199–201.

Shrikhande, M.S. (2001). Combinatorics of block designs, in *Recent Advances in Experimental Designs and Related Topics* (eds. S. Altan and J. Singh), 178–192. Nova Science Publishers, Inc., New York.

Shrikhande, S.S. (1951). On the non-existence of affine resolvable balanced incomplete block designs, *Sankhyā* **11**, 185–186.

Shrikhande, S.S. (1959). The uniqueness of the L_2 association scheme, *Ann. Math. Statist.* **30**, 781–798.

Shrikhande, S.S. (1962). On a two parameter family of balanced incomplete block designs, *Sankhyā*, **A24**, 33–40.

Shrikhande, S.S. (1973). Strongly regular graphs and symmetric 3-designs, in *A Survey of Combinatorial Theory* (eds. J.N. Srivastava *et al.*), 403–409.

Shrikhande, S.S. (1976). Affine resolvable balanced incomplete block designs: A survey, *Aeq. Math.* **14**, 251–261.

Shrikhande, S.S. and Raghavarao, D. (1963). Affine α-resolvable incomplete block designs, in *Contribution to Statistics*. Presented to Prof. P.C. Mahalanobis on the occasion of his 70th birthday, Pergamon Press, 471–480.

Shrikhande, S.S. and Singh, N.K. (1962). On a method constructing symmetrical balanced incomplete block designs, *Sankhyā* **24A**, 25–32.

Shrikhande, S.S. and Singhi, N.M. (1972). Generalised residual designs, *Utilitas Math.* **1**, 191–201.

Sia, L.L. (1977). Optimum spacings of elementary treatment contrasts in symmetrical 2-factor PA block design, *Canad. J. Statist.* **5**, 227–234.

Singh, R. and Raghavarao, D. (1975). Applications of linked block designs in successive sampling, *Applied Statistics* (ed. R.P. Gupta), North-Holland, 301–309.

Singh, R., Raghavarao, D. and Federer, W.T. (1976). Applications of higher associate class PBIB designs in multidimensional cluster sampling, *Estadistica* **30**, 202–209.

Singla, S.L. (1977a). An extension of L_2 designs, *Austr. J. Statist.* **19**, 126–131.

Singla, S.L. (1977b). A note on the construction of modified cubic designs, *J. Indian Soc. Agric. Statist.* **29**, 95–97.

Sinha, B.K. and Sinha, B.K. (1969). Comparison of relative efficiency of some classes of designs, *Calcutta Statist. Assoc. Bull.* **18**, 97–122.

Sinha, K. (1977). On the construction of n-associate PBIBD, *J. Statist. Plan. Inf.* **1**, 113–142.

Sinha, K. (1978a). A resolvable triangular partially balanced incomplete block design, *Biometrika* **65**, 665–666.

Sinha, K. (1978b). Partially balanced incomplete block designs and partially balanced weighing designs, *Ars Comb.* **6**, 91–96.

Sinha, K. and Jones, B. (1988). Further equineplicate balanced block designs with unequal block sizes, *Statist. Prob. Lett.* **6**, 229–230.

Sinha, K. and Kageyama, S. (1992). E-optimal 3-concurrence most balanced designs, *J. Statist. Plan. Inf.* **31**, 127–132.

Sinha, K. and Saha, G.M. (1979). On the construction of balanced and partially balanced ternary design, *Biom. J.* **21**, 767–772.

Sinha, K., Mathur, S.N. and Nigam, A.K. (1979). Kronecker sum of incomplete block designs, *Utilitas Math.* **16**, 157–164.

Sinha, K. Singh, M.K., Kageyama, S. and Singh, R.S. (2002). Some series of rectangular designs, *J. Statist. Plan. Inf.* **106**, 39–46.

Smith, L.L., Federer, W.T. and Raghavarao, D. (1974). A comparison of three techniques for eliciting truthful answers to sensitive questions, *Proc. Soc. Statist. Sec., Amer. Statist. Assoc.*, 447–452.

Solorzono, E. and Spurrier, J.D. (2001). Comparing more than one treatment to more than one control in incomplete blocks, *J. Statist. Plan. Inf.* **97**, 385–398.

Spence, E. (1971). Some new symmetric block designs, *J. Comb. Th.* **11A**, 229–302.

Sprott, D.A. (1954). A note on balanced incomplete block designs, *Canad. J. Math.* **6**, 341–346.

Sprott, D.A. (1962). Listing of BIB designs from $r = 16$ to 20, *Sankhyā* **24A**, 203–204.

Srivastava, J.N. and Anderson, D.A. (1970). Some basic properties of multidimensional partially balanced designs, *Ann. Math. Statist.* **41**, 1438–1445.

Stahly, G.F. (1976). A construction of PBIBD (2)'s, *J. Comb. Th.* **21A**, 250–252.

Steiner, J. (1853). Combinatorische Aufgabe, *J. für Reine und Angewandte Math.*, **45**, 181–182.

Street, A.P. and Street, D.J. (1987). *Combinatorics of Experimental Designs*, Oxford Science Publications, New York.

Stufken, J. (1987). A optimal block designs for comparing test treatments with a control, *Ann. Statist.* **15**, 1629–1638.

Stufken, J. and Kim, K. (1992). Optimal group divisible treatment designs for comparing a standard treatment with test treatments, *Utilitas Math.* **41**, 211–228.

Sukhatme, P.V. and Sukhatme, B.V. (1970). *Sampling Theory of Surveys with Applications* 2nd edn., Food and Agriculture Organization, Rome.

Surendran, P.U. (1968). Association matrices and the Kronecker product of designs, *Ann. Math. Statist.* **39**, 676–680.

Takahasi, I. (1975). A construction of 3-designs based on spread in projective geometry, *Rep. Statist. Appl. Res.* **22**, 34–40.

Takeuchi, K. (1961). On the optimality of certain type of PBIB designs, *Rep. Statist. Res. Un. Japan Sci. Eng.* **8**, 140–145.

Takeuchi, K. (1962). A table of difference sets generating balanced incomplete block designs, *Rev. Inst. Int. Statist.* **30**, 361–366.

Taylor, M.S. (1972). Computer construction of cyclic balanced incomplete block designs, *J. Statist. Comput. Simul.* **1**, 65–70.

Taylor, M.S. (1973). Computer construction of balanced incomplete block designs, *Proc. 18th Conf. Design Expt. Army Res. Dev. Testing*, 117–123.

Teirlinck, L. (1973). On the maximum number of disjoint Steiner triple systems, *Disc. Math.* **6**, 299–300.

Tharthare, S.K. (1963). Right angular designs, *Ann. Math. Statist.* **34**, 1057–1067.

Tharthare, S.K. (1965). Generalized right angular designs, *Ann. Math. Statist.* **36**, 1535–1553.

Tharthare, S.K. (1979). Symmetric regular GD designs and finite projective planes, *Utilitas Math.* **16**, 145–149.

Tjur, T. (1990). A new upper bound for the efficiency factor of a block design, *Aust. J. Statist.* **32**, 231–237.

Tocher, K.D. (1952). The design and analysis of block experiments, *J. Roy. Statist. Soc.* **14B**, 45–91.

Trail, S.M. and Weeks, D.L. (1973). Extended complete block designs generated by BIBD, *Biometrics* **29**, 565–578.

Trivedi, H.T. and Sharma, V.K. (1975). On the construction of group divisible incomplete block designs, *Canad. J. Statist.* **3**, 285–288.

Trung, T.V. (2001). Recursive constructions of 3-designs and resolvable 3- designs, *J. Statist. Plan. Inf.* **95**, 341–358.

Tukey, J.W. (1949). One degree of freedom for non-additivity, *Biometrics* **5**, 232–242.

Tyagi, B.N. (1979). On a class of variance balanced block designs, *J. Statist. Plan. Inf.* **3**, 333–336.

Vakil, A.F., Parnes, M.N. and Raghavarao, D. (1990). Group testing when every item is tested twice, *Utilitas Math.* **38**, 161–164.

Van Buggenhaut, J. (1974). On the existence of 2 designs S_2 (2, 3, v) without repeated blocks, *Disc. Math.* **8**, 105–109.

Van Lint, J.H. (1973). Block designs with repeated blocks and $(b, r, \lambda) = 1$, *J. Comb. Th.* **15A**, 288–309.

Van Lint, J. H. and Ryser, H.J. (1972). Block designs with repeated blocks, *Disc. Math.* **3**, 381–396.

Van Tilborg, H. (1976). A lower bound for v in a t-(v, k, λ) design, *Disc. Math.* **14**, 399–402.

Vanstone, S.A. and Schellenberg, P.J. (1977). A construction for BIBDS based on an intersection property, *Utilitas Math.* **11**, 313–324.

Vartak, M.N. (1955). On an application of the Kronecker product of matrices to statistical designs, *Ann. Math. Statist.* **26**, 420–438.

Veblen, O. and Bussey, W.H. (1906). Finite projective geometries, *Tran. Amer. Math. Soc.* **7**, 241–259.

Wallis, J. (1969). Two new block designs, *J. Comb. Th.* **7**, 369–370.

Wallis, J. (1973). Kronecker products and BIBDS, *J. Comb. Th.* **14A**, 248–252.

Warner, S.L. (1965). Randomized response: A survey technique for eliminating evasive answer bias, *J. Amer. Statist. Assoc.* **60**, 63–69.

Weideman, C.A. and Raghavarao, D. (1987a). Some optimum non- adaptive hypergeometric group testing designs for identifying two defectives, *J. Statist. Plan. Inf.* **16**, 55–61.

Weideman, C.A. and Raghavarao, D. (1987b). Non-adaptive hypergeometric group testing designs for identifying at most two defectives, *Comm. Statist.* **16A**, 2991–3006.

Westlake, W.J. (1974). The use of balanced incomplete block designs in comparative bioavailability trials, *Biometrics* **30**, 319–327.

Whitaker, D., Triggs, C. and John, J.A. (1990). Construction of block designs using mathematical programming, *J. Roy. Statist. Soc.* **52B**, 497–503.

Whittinghill, D.C. (1995). A note on the optimality of block designs resulting from the unavailability of scattered observations, *Utilitas Math.* **47**, 21–32.

Williams, E.R. (1975a). A new class of resolvable block designs, Ph.D. Thesis, University of Edinburgh.

Williams, E.R. (1975b). Efficiency balanced designs, *Biometrika* **62**, 686–689.

Williams, E.R. (1976). Resolvable paired comparison designs, *J. Roy. Statist. Soc.* **38B**, 171–174.

Williams, E.R. (1977a). A note on rectangular lattice designs, *Biometrics* **33**, 410–414.

Williams, E.R. (1977b). Iterative analysis of generalized lattice designs, *Aust. J. Statist.* **19**, 39–42.

Williams, E.R. and Patterson, H.D. (1977). Upper bounds for efficiency factors in block designs, *Aust. J. Statist.* **19**, 194–201.

Williams, E.R., Patterson, H.D. and John, J.A. (1976). Resolvable designs with two replications, *J. Roy. Statist. Soc.* **38B**, 296–301.

Williams, E.R., Patterson, H.D. and John, J.A. (1977). Efficient two-replicate resolvable designs, *Biometrics* **33**, 713–717.

Wilson, R.M. (1972a). An existence theory for pairwise balanced designs I. Composition theorems and morphisms, *J. Comb. Th.* **13**, 220–245.

Wilson, R.M. (1972b). An existence theory for pairwise balanced designs II. The structure of PBD-closed sets and existence conjectures, *J. Comb. Th.* **13**, 246–273.

Wilson, R.M. (1974). An existence theory for pairwise balanced designs III. Proof of existence conjectures, *J. Comb. Th.* **18**, 71–79.

Wilson, R.M. (1982). Incidence matrices for *t*-designs, *Lin. Alg. Appl.* **46**, 73–82.

Wynn, H.P. (1977). Convex sets of finite population plans, *Ann. Statist.*, **5**, 414–418.

Xia, T., Xia, M. and Seberry, J. (2003). Regular Hadamard matrix, maximum excess and SBIBD, *Aust. J. Comb.* **27**, 263–275.

Yamamoto, S., Fuji, Y. and Hamada, N. (1965). Composition of some series of association algebras. *J. Sci. Hiroshima Univ.* **29A**, 181–215.

Yates, F. (1936a). Incomplete randomized blocks, *Ann. Eugen.* **7**, 121–140.

Yates, F. (1936b). A new method for arranging variety trials involving a large number of varieties, *J. Agric. Sci.* **26**, 424–455.

Yates, F. (1940). Lattice squares, *J. Agric. Sci.* **30**, 672–687.

Yates, F. and Grundy, P.M. (1953). Selection without replacement from within strata with probability proportional to size, *J. Roy. Statist. Soc.* **15B**, 253–261.

Yeh, C.-M. (1986). Conditions for universal optimality of block designs, *Biometrika* **73**, 701–706.

Yeh, C.-M. (1988). A class of universally optional binary block designs, *J. Statist. Plan. Inf.* **18**, 355–361.

Yeh, C.-M. and Bradley, R.A. (1983). Trend free block designs: Existence and construction results, *Comm. Statist.* **12B**, 1–24.

Yeh, C.-M., Bradley, R.A. and Notz, W.I. (1985). Nearly trend free block designs, *J. Amer. Statist. Assoc.* **80**, 985–992.

Youden, W.G. (1951). Linked blocks: A new class of incomplete block designs (abstract), *Biometrics* **7**, 124.

Author Index